The Big Book of
Bags, Tags, and Labels

COLLINS DESIGN

An Imprint of HarperCollinsPublishers

First Edition published by:
maomao publications in 2009
Via Laietana, 32, 4th fl., of. 104
08003 Barcelona, Spain
Tel.: +34 93 268 80 88
Fax: +34 93 317 42 08
www.maomaopublications.com
mao@maomaopublications.com

English language edition first published in 2009 by:
Collins Design
An Imprint of HarperCollins*Publishers*,
10 East 53rd Street
New York, NY 10022
Tel.: (212) 207-7000
Fax: (212) 207-7654
collinsdesign@harpercollins.com
www.harpercollins.com

Distributed throughout the world by:
HarperCollins*Publishers*
10 East 53rd Street
New York, NY 10022
Fax: (212) 207-7654

Publisher: Paco Asensio

Editorial Coordination: Anja Llorella Oriol

Editor & Text: Cristian Campos

Translation: Francesc Zamora

Art Direction: Emma Termes Parera

Layout: Maira Purman

ISBN: 978-0-06-169171-3
Library of Congress Control Number: 2009927465

Printed in Spain
First Printing, 2009

**Never judge a book by its cover.
Or maybe you should. Let me explain**

It is common knowledge that nothing may be judged for its exterior appearance, that the content is more important than the container and that a good product does not need a good packaging so that the client recognizes its excellent qualities. Beyond the cliché which we have been bombarded with since our youth, when we used to press our noses against the window display of our first known toy store, this peculiar belief reveals a conception of reality, a vital philosophy if you prefer, that denies its human nature. It is the philosophy that has led us to this absurd aesthetic elitism that defends atonal music, windowless houses like antinuclear bunkers, subjective and incoherent literature, abstract art and other similar masochistic pleasures. The first teaching that can be extracted from this philosophy is what deals with aesthetics and what is pleasant as well as rational, needs to give way to the bizarre, disturbing and irrational. One might have in mind at this very moment a specific example of nonsensical design that fits this definition: a sofa that resembles more of an Iron Lady than a warm and soft object able to accommodate a body; a pair of shoes that seem to be designed for the claws of a platypus rather than for the feet of a human being; or, getting to the subject, a tag, work of some fanatic religious Order of Horrendous Minimalism. The second teaching of this philosophy is that people

and things enjoy a "magic essence" or "soul" that cannot be detected by means of external signs or pure aesthetics. So, let's focus on objective elements. This explains the need for (very expensive) experts in art, in politics and in psychology who unravel for us, unfortunate mortals, what we should buy, see, listen or vote.

So now you know: before you purchase a book just by reading the synopsis, the author's biography and the graphics on the cover, read it. Beginning to end. Right there and then. In the bookstore (watch for the shop clerk; the kids will get dinner ready). Then, only then you should buy it. But if you don't have time to read all the books in the store, hire an expert.

In fact, if anyone has invested a few hours of their life in designing packaging or tags for a particular product, it is possible (not certain but possible) that a proportionally similar number of hours could have been spent into the actual product. Similarly, someone who has neglected the design of the container might have also neglected that of the content. Those who are in favor of not judging by the cover will say that anyone with some knowledge about human psychology could spend a few hours on the design of the container and none to the product knowing that it is likely that all will go well thanks to our tendency to trust exterior appearances. The truth is that the probability that someone dedicates 10 hours on the container of the product and just one on its particular design is irrelevant compared with the probability that someone who has spent 10 hours thinking on the container will also spend 10 hours, if not more, to its content.

This means that the label counts.

This book features more than 300 tags and bags designed by some of the best brands and international design agencies of our time. The text has been divided into three chapters: Bags, tags and labels. Tag should be understood as any identifying or promotional element in paper, plastic, fabric or other material that provides information about the content, the origin, the composition or the use of an object to which it is attached. I understand tags as any graphic design which purpose is that mentioned above, regardless of whether it adopts the form of an object separated from the product or it has been printed on the packaging. In the chapter on labels, I have included all those identifying or promotional elements that have not been glued or stitched on the product, but rather hang from it by means of ribbons, strings, chains or any other similar solutions. The difference, in any case, is purely theoretical, given that in common practice a plain sticker glued to a particular product given as a present to a client could serve as a label for the brand, whether it swings or not.

In the back of the book, I have included a directory with the names, the websites and contact information of the more than 60 designers and design agencies included in the volume. I have also included the contact information of the brands that have their own design department.

Bags

Whether they are made of paper, plastic, cardboard or fabric, the bags included in this chapter are representations of the latest trends in the graphic design field and achieve to perfection the double commitment for which they were created: First, to appear as visually attractive as the product they contain. Secondly, to become an object of desire in and of themselves, mobile advertising panels which the consumers will carry around the streets of their city. Then they inadvertently get transformed into the brand's best commercials.

The GYG typography created by The Creative Method, the contrast between the black and white photographs and the yellow logo and the texts in Spanish reinforce the feeling of being in Mexico and provide the brand with "authenticity."

To-go bag with yellow handles. Design by Tony Ibbotson, Andy Yanto and Mayra Minobe of The Creative Method for the fast food Mexican brand Guzman y Gomez.

Bag with white handles by The Creative Method for Guzman y Gomez.

The sticker to close the to-go paper bags from Guzman y Gomez symbolizes the Mexico sun. A vivid yellow was selected in order to achieve the greatest possible contrast and visual impact.

Paper bag for nachos by Tony Ibbotson, Andy Yanto and Mayra Minobe of The Creative Method for Guzman y Gomez.

THE COUNTR...
SELECTED TH...
CUTS OF BEEF, LEA...
AND SKINLESS FREE
.RANGE CHICKENS. WE
USE ONLY OCEAN FRESH
SEAFOOD, FRESH GARDEN
VEGETABLES AND ARE
EQUALLY STRICT WITH
ALL OUR SEASONINGS.
THIS IS FOOD
TRADITIONALLY HIGH
IN NUTRIENTS, PACKED
WITH VITAMINS AND
LOW IN FAT. WITH ONLY
THE EXCEPTION OF
OUR TOTOPOS, NOTHING
IS COOKED IN OIL.

Auténtico, Mexicano, Delicioso

GUZMAN Y GOMEZ *Taquería*

To-go bag and closing sticker designed by The Creative Method for Guzman y Gomez.

In summer 2005, the six lines of the French brand KanaBeach were reorganized into three: Biologik (natural materials), Circus (urban fashion) and Elegant (night). The three bags on this page have been designed as promotional elements of the three lines.

Promotional fabric bags designed by KanaBeach.

Paper bag by KanaBeach. Available in two different sizes at the brand's stores.

Th' Rezidents is the name of a line of 13 characters designed for the Japanese brand of urban fashion Ooito Japan. The character printed on the bag can be cut out and made into a rag doll. The instructions are printed on the bag.

Cotton bag designed by Sonia Chow & Huschang Pourian
of ChowPourianLab for Ooito Japan (front and back).

The graphic pattern of the Green Bag pays homage to the gundam Japanese culture. The Blue Bag was designed during the 2006 soccer World Cup and shows a PlayStation remote control, the only scenario where the Japanese soccer team "can beat" Spain.

Green Bag (left) and Blue Bag (right). Design by Sonia Chow & Huschang Pourian of ChowPourianLab for Ooito Japan.

The bags for the shirts of the brand Ooito Japan are decorated with reinterpreted and modernized icons of the Japanese tradition. The goal is to create a bridge between the Japanese contemporary pop culture and its roots.

Plastic bags for T-shirts designed by Sonia Chow & Huschang Pourian of ChowPourianLab for Ooito Japan.

Detail of the graphics that decorate the bags for the Ooito Japan T-shirts.

Marjolein Delhaas chose to print the list of the products for sale (using ITC Avant Garde Gothic font), on the front and back of the Foodelicious bags rather than the logo, which was relegated to the sides of the bags. The goal is to turn the bag into a gift-like object with a powerful and modern image.

Paper bag designed by Marjolein Delhaas for the food and gift store Foodelicious in Rotterdam.

The typeface used here—Mari, designed by the agency e-Types—conveys simplicity. The bag was used in the presentation of the male spring-summer 2009 collection of the Danish fashion house Mads Nørgaard-Copenhagen.

Promotional plastic bag designed by the Danish firm Mads Nørgaard.

The plastic bag for Byggfabriken, a Swedish company that specializes in fabrication and sales of materials for the restoration of period interiors, shows rows of iconic characters doing the tasks associated with such restorations. The font is Futura.

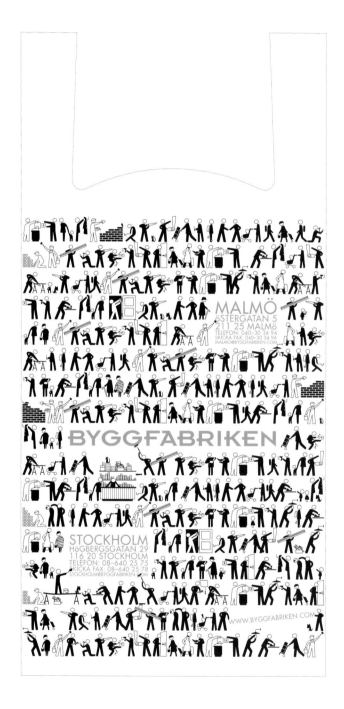

Design by Dan Jonsson and Louise Hederström of Charming Unit for Biggfabriken.

Top: self-closing plastic bags designed by Dan Jonsson and Louise Hederström of Charming Unit for Byggfabriken.

Available in two colors and in different sizes, the gigantic unisex bag Packrat for Lomographic Society International has been decorated with the brand's original Russian logo and with one of its first cameras, the Lomo LC-A, framed in a laurel wreath.

Packrat cotton bag with interior pocket and metallic buttons designed for Lomographic Society International.

Cardboard boxes designed for the Packrat bag. Large and extra-large.

On this page: the promotional paper bags for Lomo adopt the form of the original cameras of the Russian brand. On the next page: the waterproof plastic bag designed for the Lomo camera with fisheye lens can be used as a beach tote bag.

Promotional paper bags designed by Lomographic Society International.

Plastic bags by Lomographic Society International designed
for the Fisheye Underwater Combo cameras.

Every year, the Spanish shoe brand Camper produces a special edition of bags for Christmas and New Year's. The two bags on this page correspond to the 2005 edition. Those on the next page, with the numbers 25 (Christmas) and 01 (New Year's) printed, correspond to the 2006 edition.

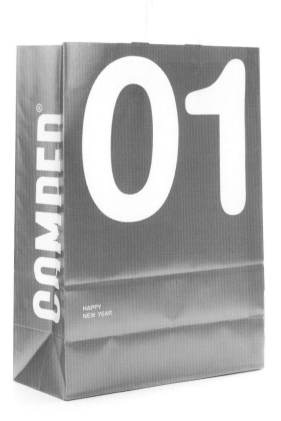

Special edition of paper bags designed by Pablo Martín and Bárbara
Castro of grafica for Camper.

Holding the promotional bag designed for the restaurant Kompot in Odessa (Russia) that specializes in home cooking, the user seems to carry a net bag with a jam jar, a playful wink to Kompot´s home cooking culinary philosophy.

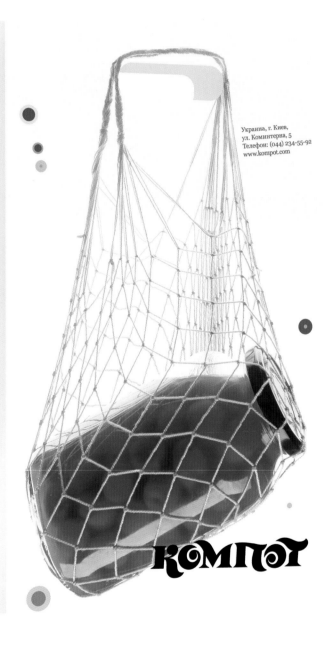

Украина, г. Киев,
ул. Коминтерна, 5
Телефон: (044) 234-55-92
www.kompot.com

КОМПОТ

Design by Anton Schnaider from Art. Lebedev Studio for Kompot.
Typography designed by Zakhar Yashchin.

The logos and the graphics that grace the bags of the Swiss brand +alprausch+ are exactly the same visuals used on their postcards, vinyl banners, and other promotional brand materials.

Plastic and paper bags designed by Thomas Brader of Komun GmbH
for the Swiss brand +alprausch+.

The fashion and design boutique Romeo & Juliet Couture commissioned a modern, emotional and international logo that would reflect elements of the traditional Hungarian art. Keeping this in mind, the typography chosen, Folk Art Sampler, reminds one of the pixilated typographies used in the eighties.

Fabric bag. Design by David Barath in collaboration with Visual Group for Romeo & Juliet Couture.

Line of cardboard bags in different sizes designed by David Barath
in collaboration with Visual Group for Romeo & Juliet Couture.

While black was reserved for the Deluxe line bags of the brand, red, white and yellow, typical colors in traditional Hungarian art, were used for the bags and the promotional elements of Romeo & Juliet Couture fashion casual line.

Logo designed by David Barath in collaboration with Visual Group for Romeo & Juliet Couture.

Promotional bag for the casual line of Romeo & Juliet Couture.

Designed for the spring-summer 2009 collection of the line Lonely Tiger, supposedly based on a Russian traditional legend about the Amur Tiger, the saddest in the world. This bag and all the materials have been entirely fabricated in Finland.

Cotton bag included in the press kit of IVANAhelsinki. Design by Paola Ivana Suhonen.

To emphasize and accentuate the personality of the Dutch skate store Bonk, Staynice designed a graphic image based on the Helvetica font. The colors are Pantone 3005 C, Process Black C, Cool Gray 10 C, and Cool Gray 8 C.

Plastic bag designed by Staynice for the skater material store Bonk.

The different sizes of the bags designed by grafica for the Spanish fashion brand Gimenez&Zuazo differ from each other in the vivid color of the interior lining that matches that of the bag's fabric handle and the logo's initials of the brand.

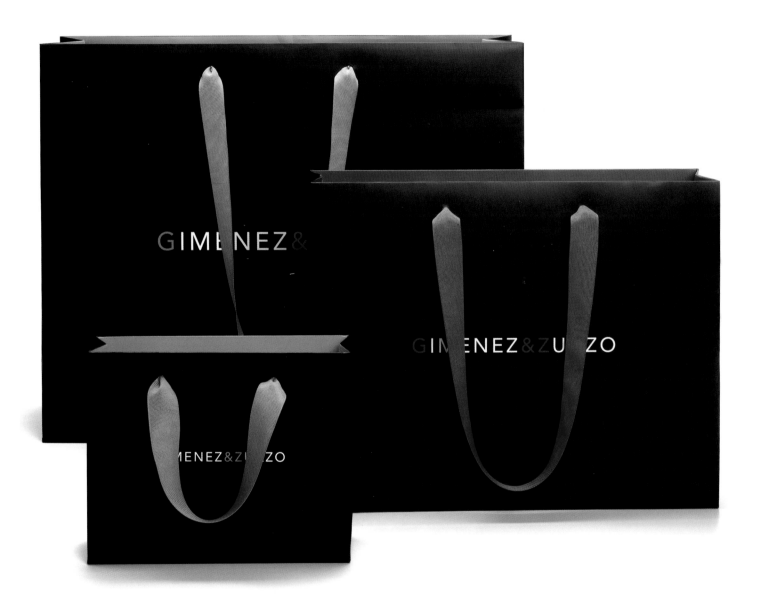

Bags designed by Pablo Martín and Meri Lannuzzi of grafica for Gimenez&Zuazo.

Test Tube is a store in Perth (Australia) specializing in "extraordinarily extraordinary" design objects. The black on the bag contrasts with the silver finish of the logo and imparts a sense of elegance.

Bag designed by Mark Braddock and Nic Bowen-Sant from Block for Test Tube, a store of design objects (Perth, Australia).

A logo with a digital aesthetic for the PL-Line store is the axis around which revolve all the elements designed by Stormhand, from the bags to the website. The logo, with a futuristic aesthetic, contrasts with the classic and exclusive style of the boutique.

Design by Boy Bastiaens of Stormhand for PL-Line.

On the front of the boxes by Atelier LaDurance is a photograph of a metallic capsule. The capsule also serves as "repair" kit. The capsule is transformed in the brand's image and is its most powerful visual identifier.

Cardboard box designed by Boy Bastiaens of Stormhand for Atelier LaDurance.

The icon for Conflict, a store specializing in Dutch furniture and design objects is a medieval castle surrounded by storm dark clouds, a design of ominous portent that pays homage to the symbols in traditional heraldry.

Conflict Designmarket / Minckelersstraat 14 / 6211gx / Maastricht / www.conflict.to

Design by Boy Bastiaens of Stormhand for Conflict.

Designed for the store in London's Soho district, this Lazy Oaf bag was made out of pink recycled plastic—pink being the predominant color in the store, along with black.

Pink plastic bag designed by Gemma Shiel for Lazy Oaf.

Two simple embroidered geometric elements and the corporate type font are sufficient to decorate one of the fabric bags designed by SKUNKFUNK, a design that has survived the ups and downs and the passing trends on the shelves of this Spanish brand.

Embroidered bag designed by Susana Sánchez Monje and Sergio Llanos of SKUNKFUNK.

In the frame of the Art-Push program, from the collaboration between the Spanish fashion brand SKUNKFUNK and the artist elDimitry, comes the design of this collection of bags decorated with the image of different wild animals in fluorescent colors and executed with fabric scraps.

Fabric bags designed by Susana Sánchez Monje and Sergio Llanos
of SKUNKFUNK in collaboration with elDimitry.

The graphics of the recycled paper bag of SKUNK-FUNK for the 2006-2008 seasons are the same as those on the brand's clothes. The logo "People Do Funk" intends to reinforce amongst the clients, the idea of belonging to a community.

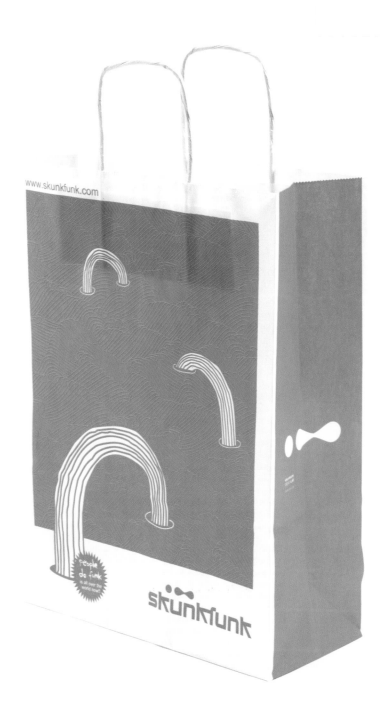

Paper bag designed by Susana Sánchez Monje and Sergio Llanos of SKUNKFUNK.

The logo of the brand Yes No Maybe painted with spray paint that can be seen on the boxes reflects the exclusiveness of its products. The boxes were removed from the market when it was observed that most of them would "get lost" when put in the mail.

Cardboard box designed by Ben Farleigh of Yes No Maybe. Hand-painted.

Paper bag with handles designed by Ben Farleigh of Yes No Maybe.

The bags for Paula Cahen d'Anvers show characters and games related with the maritime world: instructions to make a mariner's knot, a paper boat or nautical semaphore flags that can be cut out of the bag.

Graphic pattern by Fbdi for the bags of the spring-summer 2009 children's fashion line of Paula Cahen d'Anvers.

Cardboard bags by Fbdi for Paula Cahen d'Anvers.

The bags of the brand Jackie Smith, on 280 gram cardstock with matt polypropylene, show the logo printed in high relief. The handles are made of Gross ribbon. Its interior is decorated with a pattern of shoes, purses and other accessories.

Line of bags and boxes designed by Fbdi for the brand Jackie Smith.

Fabricated with 216-gram paper and with matte polypropylene on its front and sectorized lacquer over the logo, the bag for the brand Paula Cahen d'Anvers has a black interior with UV lacquer. The cord was specially designed for the brand.

Bags and labels designed by Fdbi for the women's line of the brand
Paula Cahen d'Anvers.

Using one ink printed on coated paper, this bag is inspired on the engravings of the silver collection of the Spanish jewelry brand joid'art. The manual gesture of the design is actually the result of the scanning of the jewelry and the manipulation of the image.

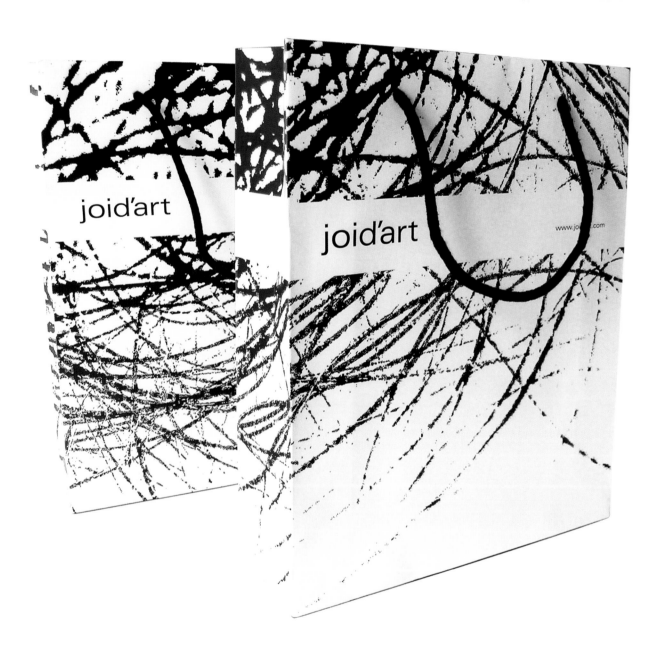

Bag designed by Cristina Julià and the creative studio of the company for the contemporary silver jewelry shops joid'art.

Taking into account the specifications previously established for the typo font and the colors (black, brown and earth colors) of Anasol's corporate image, the design makes the most of this bag. This explains the use of all four sides as graphic surfaces.

Development of the bag designed by Nicolás Díaz of Plasma
for the leather accessories brand Anasol.

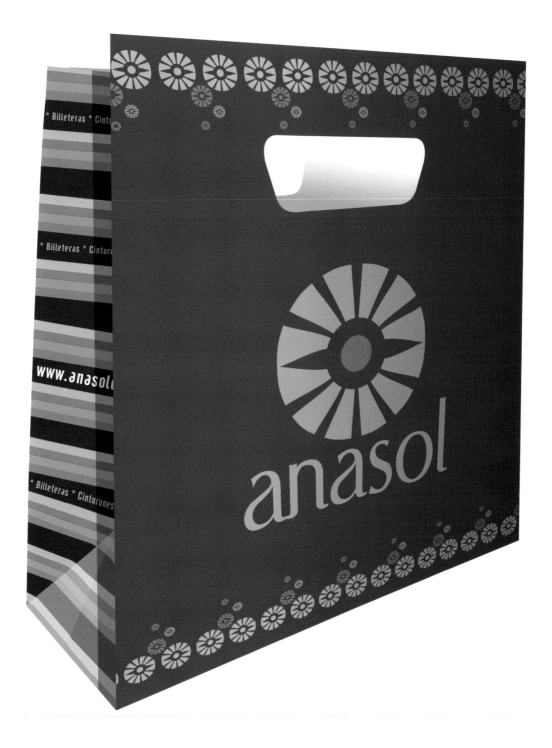

Rendering of the bag designed by Plasma for Anasol.

In addition to the brand's logo, and like all other stationery products, the bag for Bo van Melskens shows mysterious disjointed sentences without an apparent meaning, but with a strong evocative potential that aims to wake the client's curiosity.

Bag designed by Stefanie Reeb for the fashion brand Bo van Melskens.

Development of the promotional bags for Kulte. One of them has been ironically decorated with bundles of bills, creating a pattern that is easily identifiable at first sight by the clients and people familiar with the brand.

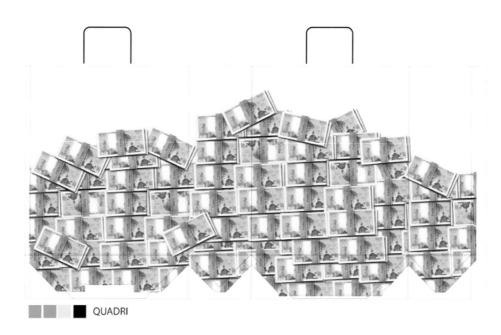

QUADRI

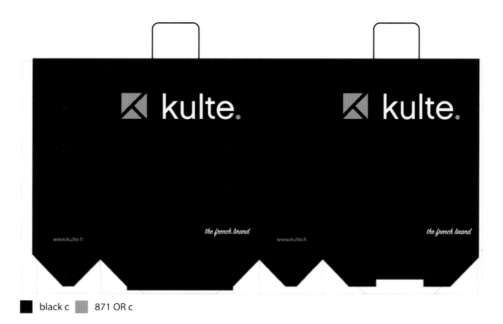

black c   871 OR c

Bags designed by Yak da House, de Kulte, for the Let's Dance collection.

lsb

Faldas de Autor

lsbfaldasdeautor@hotmail.com

Tags

Swinging tags are, in a way, the "ultimate" tags, those preferred by designers. While the design of conventional tags is restricted simply because they are destined to be sown, glued or printed on the packaging or the product, swinging tags offer designers the greatest leeway. This explains double, transparent, and superimposed tags, the use of unconventional materials, or baroque and extreme designs. More power to fantasy.

Onda's tags, of slightly irregular geometry, hang from a ribbon attached to the garments on which the brand's logo has been printed. The models in the photographs are representative of the Portuguese brand's typical target.

Tags designed by Onda's creative team (front and back).

The colors on Onda's tags help the client identify the different lines of the brand. The web page address is printed both on the front and on the back of the tag, which also presents an abstract design based on the company's logo.

Tag designed by Onda (front and back).

Onda's men and women's lines have different tags. The colors used for the text (brown for the unisex or men's lines, and orange for the women's lines) and for the brand's logo are not the same. Neither are the graphic patterns.

Folding tag designed by Onda.

Onda designs for the men's (left) and women's (right) collection.

Onda is a brand specialized in neoprene thermal garments for aquatic athletes. This explains the graphic patterns used in many of the tags expressing the surf and kitesurf aesthetic.

Tag designed by Onda (front and back).

Tags

Faded edges, off-whites, eighties typography and comic-like illustrations are the visuals around which revolve the designs for the Kulte spring-summer 2009 Beach Party tags.

Tag designed by Malax Design for Kulte.

Design by Malax Design for Kulte.

The typography and the black color on the tags for Alpinestars' Nero line were chosen for their suitability to minimalist and sophisticated air of the garments to which they are attached, breaking away from the brand's traditionally colorful and sporty image.

Tags for Alpinestars' main line of clothes and the Nero line by Alpinestars' design team.

Tags designed by Alpinestars.

Tags

The tag for the main line (center), the Japanese vintage jeans and the Multifunction Track Jacket (right) was designed starting from the Sans font and from Squad's official colors (black and red).

Tags designed by KGB+FOX for Squad's jeans.

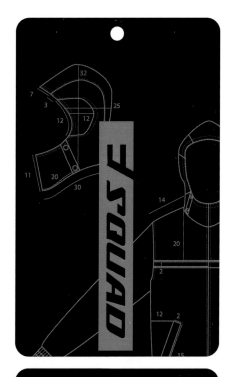

JAPAN DENIM

Genuine Motorcycle Spirit

esquad-jeans.com
ESQUAD

Retour aux sources. Pour fabriquer le jean premium JAPAN DENIM, Esquad® est allé rechercher au Japon les métiers à tisser utilisés jadis par les pionniers du jean, les seules machines capables de donner à la toile sa vraie densité. Tissé en petite laize dans le respect des techniques originelles, sur des métiers à navettes et avec un fil retors, le JAPAN DENIM offre une résistance et une longévité aujourd'hui oubliées. Cette exceptionnelle solidité a fait du jean l'équipement incontournable des premiers motards américains. Esquad® l'a retrouvée.

Back to the roots. To produce JAPAN DENIM premium jeans, Esquad® looked to Japan to seek out the ancient looms used by jeans pioneers. The only machines capable of weaving such high density fabrics. Woven in narrow width, respecting original techniques with shuttle looms and twisted yarns, JAPAN DENIM is a return to long forgotten standards in strength and durability. The sturdiness that turned jeans into the staple wear for the first American bikers. Recaptured by Esquad®.

Une lisière rouge et or signe l'authenticité et la haute qualité du jean JAPAN DENIM.

A red and gold selvedge is the signature of JAPAN DENIM jeans' authenticity and high quality.

Le JAPAN DENIM offre une résistance incomparable par rapport aux jeans fabriqués selon les méthodes industrielles, mais il ne contient pas d'Armalith®. Pour une protection maximale, Esquad® recommande le choix d'un modèle en Armalith®.

JAPAN DENIM offers unrivalled resistance compared to jeans manufactured by industrial methods, but does not contain Armalith®. For maximum protection, Esquad® recommends opting for an Armalith® model.

Paris, 2003. Professionnel de la mode, P.H. Servajean rêve de piloter sa moto sans perdre son style ni sacrifier sa sécurité. Fou de technologie, le créateur français a l'idée de mélanger du coton et des fibres techniques pour créer la toile de jean la plus résistante du monde. C'est la naissance de l'Armalith® et de la marque Esquad®.

Pour prouver la performance de l'Armalith®, Esquad® s'associe aux plus grands cascadeurs. Ensemble, ils suspendent un 4x4 Hummer® à un jean et organisent une chute moto à 110 km/h. La toile résiste et Esquad® entre dans la légende du jean.

Aujourd'hui, le Laboratoire Esquad® imagine de nouvelles collections de jeans premium et Armalith® et de vêtements aux matériaux d'avant-garde.

Paris, 2003. Fashion pro P.H. Servajean dreams of riding his motorbike with no skimping on style OR safety... Technophile, the French designer conjures up a cotton/technical fibre fusion to create the most resistant jean fabric in the world. It's the birth of Armalith® and the Esquad® brand.

Top stuntmen help Esquad® prove Armalith®'s performance. Together they suspend a 4x4 Hummer® from a pair of jeans, and are pulled across the floor at 110km/h. The jeans resist... a legend is born.

Today new collections of luxury and Armalith® jeans, plus apparel made of avant-garde material are brought to you, straight out of the Esquad® lab's psyche.

esquad-jeans.com
ESQUAD
THE LEGENDARY ARMALITH® APPAREL

DONT WALK

STRONG CHOICE !

MULTIFUNCTION

TRACK JACKET

esquad-jeans.com
ESQUAD

 THERMOREGULATION //////////////////////////////////////
Équipé d'une membrane biocéramique, le Multifunction Track Jacket exploite le rayonnement infrarouge du corps pour réguler son pouvoir calorifique. Sous un blouson ou sur un t-shirt, il offre un confort thermique idéal par tous les climats, chauds ou froids. // *Doted with a bioceramic membrane, this Multifunction Track Jacket uses the infra-red rays emitted by the body to regulate heat generation. Under a coat or over a t-shirt, it allows an optimal level of thermal comfort to be achieved, in all hot or cold climate conditions.*

 HI-VISIBILITY //////////////////////////////////////
En cas d'urgence, avec sa doublure réversible fluo norme EN471, le Multifunction Track Jacket se transforme en gilet de sécurité haute visibilité. // *In case of emergency, this Multifunction Track Jacket can be converted into a hi visibility safety jacket, thanks to its reversible EN471 standard fluo lining.*

 RAINBLOCK //////////////////////////////////////
Grâce à sa membrane biocéramique, le Multifunction Track Jacket est à la fois respirant et parfaitement étanche. // *Thanks to its bioceramic membrane, this Multifunction Track Jacket provides ultimate rain protection, while simultaneously allowing the skin to breathe.*

 WINDBLOCK //////////////////////////////////////
Le Multifunction Track Jacket est entièrement coupe-vent. En moto, sa puissance de thermorégulation reste intacte, même à haute vitesse et dans les vents froids. // *This Multifunction Track Jacket is an entirely effective wind blocker. Its thermoregulation capacity remains intact even when motorcycling at high speeds and in cold winds.*

Squad's tags (front and back).

Died-cut irregularly and conceived as if it were a time-faded hunting license, the tag for the North American brand Freshjive is fringed with a camouflage fabric similar to military and hunting garments.

1. The literal inerrancy of the Scriptures
2. The second coming of Jesus Christ
3. The virgin birth (not the Immaculate Conception)
4. The physical resurrection of the body
5. The substitutionary atonement
6. The total depravity of man - original sin

GOOD UNTIL JULY 92

Ah! I knew you would come. I told my friends that you might give one or more of them a home! They hope you will. If you decide to adopt one for your page you get to name it yourself!

www.freshjive.com

Tag designed for Freshjive (front and back).

Freshjive chose brass knuckles as the distinctive element of its folding tag. The golden inside of the tag is reflected on the small brass knuckles which hangs off a string from the tag.

Tag and pendant designed by Freshjive.

Using one single bright color (red), the tags for BB Dakota achieve a hugely superior impact than those which feature multi-color designs, thanks in part to the contrast with the hand drawn logo.

Tag designed by BB Dakota.

The fashion brand BB Dakota invites its clients to become models by means of a message printed on the tags. The sentences written below the Polaroid format photos quote typical commentaries from stylists and fashion photographs.

Tag designed by BB Dakota.

Designs by BB Dakota (front and back).

Tags

A small velvet ribbon, made into a bow, allows this heart shaped fabric tag to hang from the garment. The brand's logo, the name of the line to which it belongs (Jack) and the checked pattern in the background have been embroidered on the tag.

Tag designed by BB Dakota.

Adding a small mirror to the tag —which is decorated with heraldic motifs inspired in the Middle Ages—BB Dakota makes fun of the vanity and glamour associated with the fashion world.

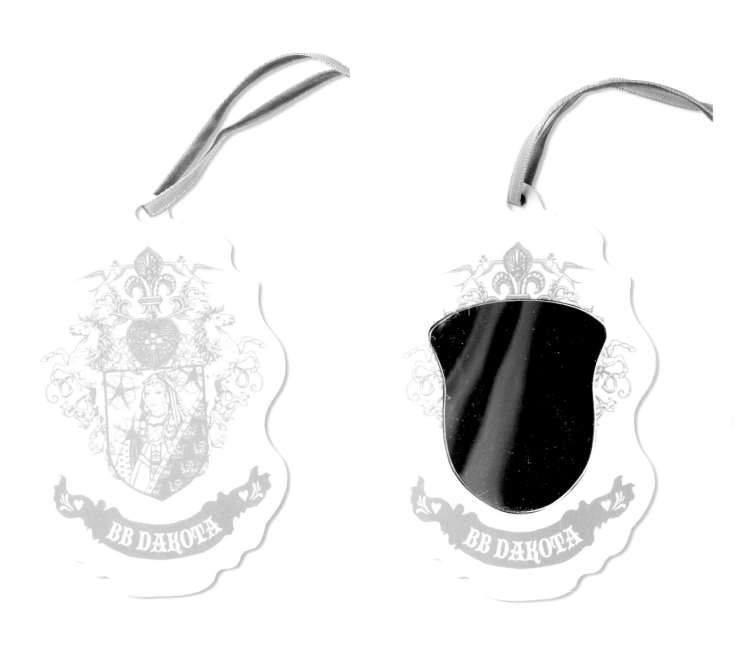

Designs by BB Dakota (front and back).

The double tags let the designers play with transparencies and graphics. In this case, BB Dakota has jointed two tags with different opacity levels and has opted for a graphic design reminiscent of something romantic yet scruffy.

Superimposed double tag designed by BB Dakota.

Tags

In this tag, the brand's logo has been die-cut in like old computer processor punch cards, perforated to record binary code. Its futuristic and minimalistic aesthetic contrasts with the romantic concept of all the other BB Dakota tags.

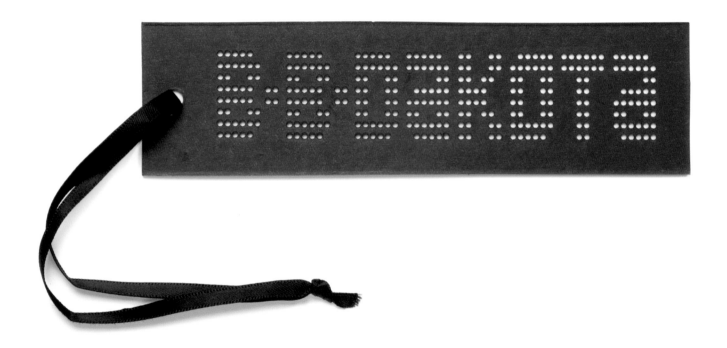

Tag design by BB Dakota.

A romantic pattern, a ribbon decorated with the hand-drawn logo of the brand and a replacement button for the garment are the basic elements of this BB Dakota tag.

Design by BB Dakota (front and back).

The padlock is one of Pepe Jeans London's key identity elements. It needs to be broken before taking it off the garment. A gigantic padlock built to look like the one on this page was part of an advertising campaign with the slogan "Read the Small Print."

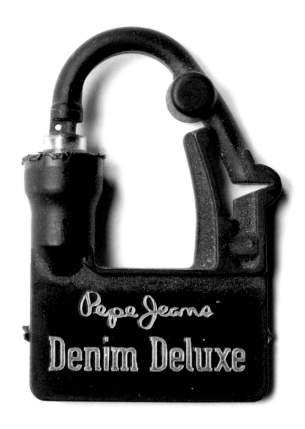

Design by Boy Bastiaens of Stormhand for Pepe Jeans London's Denim Deluxe line.

The Camargue cross is the emblem of the church of Saints-Maries-de-la-Mer. Atelier LaDurance adopts it as a graphic symbol in red, white, and blue for the padded hangers used for sweatshirts, jackets, and wool garments.

Design by Boy Bastiaens of Stormhand for Atelier LaDurance.

Atelier LaDurance's mending kit takes the form of a metallic capsule. The various elements in the kit (a thimble, two buttons and two pieces of fabric) have been inserted manually. The color of the fabric pieces is different for women and men lines.

Metallic capsule designed by Boy Bastiaens of Stormhand for Atelier LaDurance.

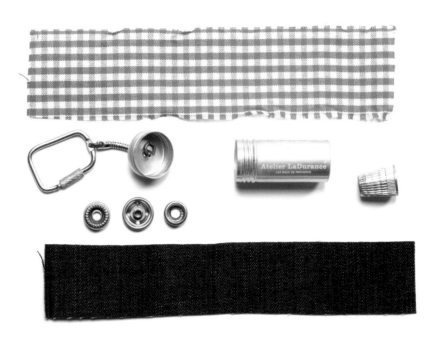

Content of the capsule for Atelier LaDurance.

Atelier LaDurance uses Japanese fabrics in some of
its exclusive lines. The tags designed for this collection show the weight in ounces (oz) of these fabrics.
For each different weight gets its own color tag.

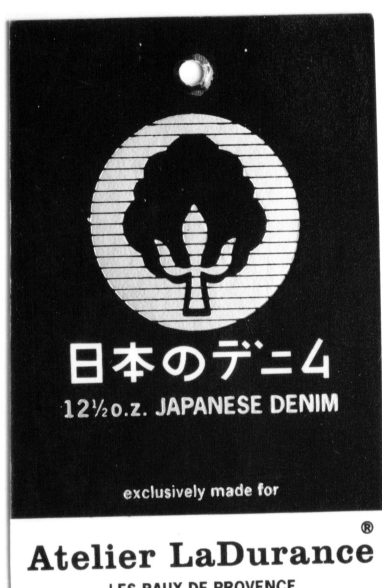

Tags designed by Boy Bastiaens of Stormhand for Atelier LaDurance.

The tag for Industry (Clothing) spring 2008 season was designed with the idea of reusing it, which explains the hole in one of its ends that reminds one of the common "Do Not Disturb" doorknob hangers in hotels.

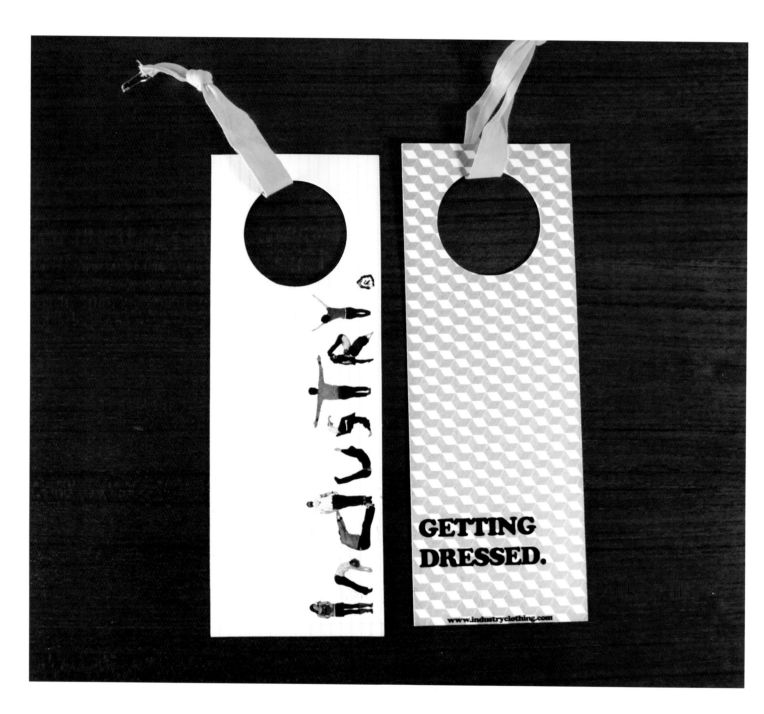

Design by Evan Melnyk of Curse of the Multiples for Industry (Clothing).

Tags designed by Evan Melnyk of Curse of the Multiples for Industry (Clothing).

Tags

The tags for Me(ël)'s fall 2008 collection are made of paper with a parchment texture and display a watercolor-like print as well as vectorial geometric designs inspired by flora.

Tag designed by Evan Melnyk of Curse of the Multiples for Me(ël).

Design by Evan Melnyk of Curse of the Multiples, for Me(ël) (front and back).

For the new Birgit Israel outlet, Richard Ardagh has designed a Rococo frieze that frames the boutique's logo, and a color system that differentiates between the contemporary and vintage garments, as well as between those produced by the same brand.

Box and tags designed by Richard Ardagh for Birgit Israel's fashion and accessories store in Chelsea (London).

Details of tags for Birgit Israel.

With the goal of fusing ecological and organic concepts with contemporary design, Haus of M has designed this tube made of French oak which contains the parchment tag for the garment rolled up inside.

Tube containing the tag designed by Simplicio Michael Luis of Haus of M for M… TheMovement.

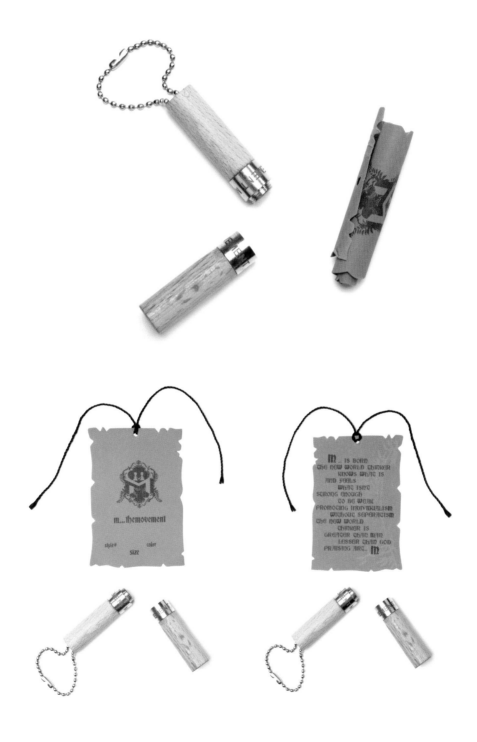

Content of the tube for tags for M… TheMovement.

Khaki leather tag, and medieval heraldry aesthetic designed for the denim garments and the jackets for the Los Angeles brand M… TheMovement. The belt for the tag is also leather.

Tag designed by Simplicio Michael Luis of Haus of M
for M... TheMovement (front and back).

Tag designed for M… TheMovement's jeans collection. All of the fonts and graphics used in the range of tags have been created specifically by Haus of M for the brand.

Designs by Simplicio Michael Luis of Haus of M for M… TheMovement.

Tags

Haus of M has designed the tags for M… The-Movement's Denia collection starting from design patterns on traditional textiles that can be found in the National Museum of Archaeology, Anthropology and History of Peru.

Tags and stickers by Simplicio Michael Luis of Haus of M for M… TheMovement.

PREMIUM CLOTHING   SIZE

100% cotton
made in China

www.MTHEMOVEMENT.com

style: _____
size: _____
color: _____
fabric: _____

_____

WWW.MTHEMOVEMENT.COM

FIND YOURSELF...FIND YOURSELF...FIND YOURSELF...

Designs by Simplicio Michael Luis of Haus of M for M… TheMovement.

The holes of these decorative metallic patches with a worn out finish permit the attachment to the garments. The M... TheMovement logo and the promotional slogans have been engraved on the metal.

Metal medals designed by Simplicio Michael Luis of Haus of M
for M... The Movement.

The double tags for +alprausch+ are decorated with "alpine" animals, such as deer and cows, in an ironic reference to its Swiss origin. The brand's logo in white over a dark blue background completes the design.

Tags designed by Thomas Brader of Komun GmbH for +alprausch+.

The illustrations on the tags designed for the brand Cosita Linda pay homage to the sensuality, joy and femininity of the latina woman, and mixes natural elements with pinup images of the fifties.

Tag designed by Felipe Ramírez of Plasma for Cosita Linda.

Design by Felipe Ramírez of Plasma for Cosita Linda.

Tags

The illustrations on the tags that can be seen on this double page have been hand drawn. Later, a worn out finish was applied representing the hand-crafted process of the fabrication of the LSB skirts.

Tags designed by Julián Román of Plasma for LSB.

The logo for the German brand of hand bags Zwei, which uses Lubalin Graph font and is printed on all the tags of the brand has been designed to be read normally or upside down (Zwei-lam2).

Tags designed by Zwei.

All the tags for Smiley Collection are made of TCF paper which does not contain chlorine. Given that the paste has not been whitened with any chlorine substance, the TCF allows a greater printing quality. Also, it is much less polluting than other types of paper.

This Smiley product will
change your life.
It has been made
with love
to remind you
how powerful a smile is.
Smiles are free...
Don't save them.
Start now
life will smile back at you!
and remember peace
begins with a smile!

Smiley World Association®
Share your smile with those in need
www.smileyworldassociation.org

the original · trademark ·
TCF Paper (100% Chlorine Free)
Smiley® collection
www.smileycollection.net

Tag designed by Kyung Park of The Smiley Company for Smiley Collection.

The simple square tag designed for Bo van Melskens shows, on the front, a slogan and the brand's logo, while the back shows a pattern made using the logo, the website, and the name of the garment's collection.

Tags designed by Stefanie Reeb for Bo van Melskens.

The tags for the fashion brand Fujizaki show, amongst an amalgam of graphic elements of various sources and aesthetics, the brand's logo designed with different fonts as well as images of its web page.

Tags designed for Fujizaki.

The contrast between the black on the front and bright orange on the reverse in reference to the name of the collection (Orange Line) is all this tag needs to capture the client's attention.

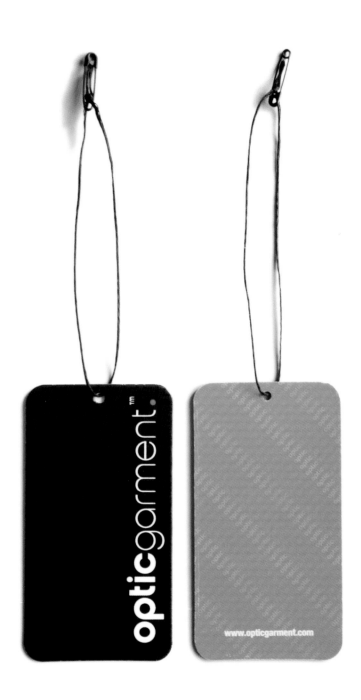

Tags designed for HakGraphics and the design department at Optic Garment for the Spanish fashion brand Optic Garment (front and back).

The 13 characters of Th' Rezidents were created as if they were real people with their own e-mail addresses. These can be found on the garment tags for the brand Ooito Japan, along with designer information.

Tag designed by Sonia Chow & Huschang Pourian
of ChowPourianLab for Ooito Japan (front and back).

Inside this origami hexagon are various Japanese pop culture icons. A small plastic piece was added to the tag to show the web page, the logo and other graphic elements of the brand.

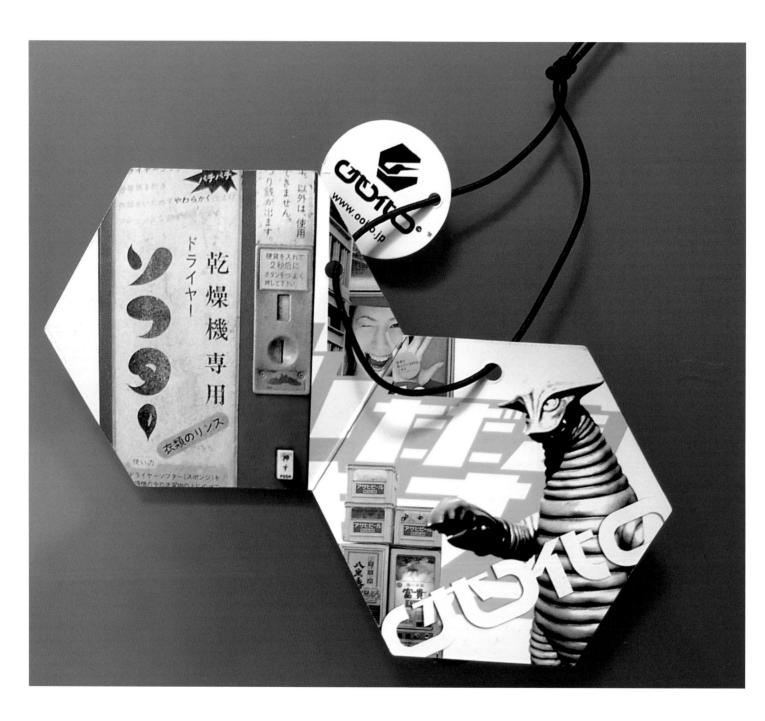

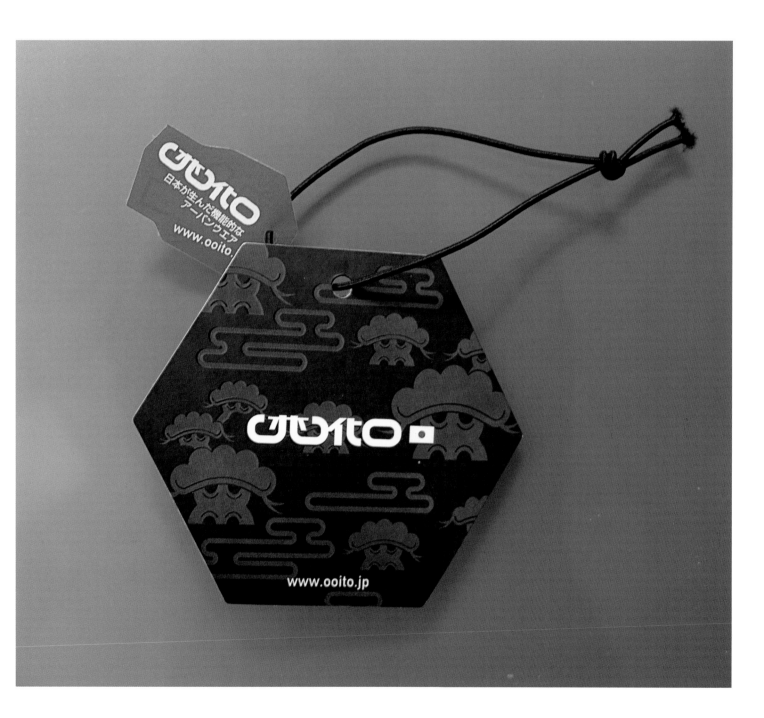

Origami tags by Sonia Chow & Huschang Pourian of ChowPourianLab for Ooito Japan.

The concept for the tags for the Lomo's Parisian
store came about from various influences: planes,
hotels and travel. This explains the designs that imitate doorknob hangers at hotels' guest rooms and
the control tags for the luggage used in airports.

Ne pas jeter sur la voie publique

LOMOGRAPHY CHECK-IN!
6, PLACE FRANZ LISZT, 75010 PARIS I HORAIRES DU MARDI À SAMEDI DU
11:00 - 19:00, JEUDI NOCTURNE JUSQU'À 21:00

ET EN DECEMBRE I DIMANCHE 14:00 - 19:00

LE NOUVEAU LOMOGRAPHY SHOP PARIS
LOMOGRAPHY CHECK-IN!

Lomography se dirige vers la capitale française et pose son propre quartier
général dans la ville de Paris. C'est dans une boutique charmante, décrochée
du temps où il n'y avait pas de style, que Lomography met ses bagages et on
ne plaisante pas avec ses affaires ! Mais il s'agit aussi de fun et de jeux.

Pour une période limitée seulement, Lomography va : éliminer son stock de
valises !

Le Check-In! de Lomography sera une « shopping experience » et un point de
rencontre pour les Lomographistes de Paris et pour tous les gens captivés
par l'incroyable monde de la photographie analogue et par les activités de la
plate-forme mondiale de Lomography. Bloquée dans la paperasserie, dans
la bureaucratie parisienne et dans les idées Lomographiques continuelles, le
design de l'espace est fait autour du concept de la transition. Le magasin
présente aussi le plus grand Mur Lomo en Europe, avec le thème du voyage,
conçu presque entièrement de vielles valises et offrant une sélection
premium de produits Lomography! Le Check-In Lomography sera aussi le
pays des événements Lomography réguliers et les workshops autour de la
photographie analogue, des instruments et des astuces lomographiques, du
charme du film analogue et de la beauté de cet engin spatial, la Terre.

Faites vos bagages et préparez-vous pour l'embarquement; vous êtes
chaleureusement invités à être parmi les premiers à essayer ce qui es en
route de devenir le premier Lomography Shop à Paris au printemps 2008 !

WWW.LOMOGRAPHY.COM I PARIS@LOMOGRAPHY.COM **lomography**

Lomographic Society International promotional varnished cardboard
tag for the Paris store.

FISHEYE NO.2
HG 3642
THE MUCH UPGRADED VERSION OF THE STANDARD FISHEYE. NOW INCLUDING MULTIPLE AND LONG-TIME EXPOSURE BUTTONS, A HOT-SHOE FOR AN ADDITIONAL FLASH AND A TRUE FISHEYE-VIEWFINDER.

ACTION SAMPLER FLASH
HG 2347
UN SEUL APPUI SUR LE DÉCLENCHEUR = SÉQUENCE DE QUATRE IMAGES AVEC UN FLASH EN QUATRE COUPS SYNCHRONISÉS. CAPTUREZ DES IMAGES SUPER JOUR ET NUIT!

LC-A+
HG 9139
THE ORIGINAL AND CLASSIC "SNAP-SHOT CAMERA IS REBORN. IN 2006 WORLD-FAMOUS FOR PROFRADINOV'S UNIQUE LENS AND THE SENSATIONALLY GLOWING COLOR IT BRINGS ON TO EVERY REGULAR 35MM FILM.

FISHEYE CAMERA
HG 3541
ITS UNIQUE PREMIUM-QUALITY GLASS FISHEYE LENS CAPTURES AN ENORMOUS 170° FIELD OF VISION AND YIELDS A NEARLY CIRCULAR IMAGE ON A STANDARD PHOTO PRINT.

Informative tags for Lomographic Society International Paris store.

The tag designed for the high end men's fashion boutique Zekka is 2 × 5 inches and has been printed in one color on thick cardstock. A red hand-sown thread detail contributes a distinctive touch.

Tag designed by Block for Zekka (front and back).

The magenta tag for Mishka is used for this New York brand's T-shirts. The untidily graffitied bear, which decorates the tag is a wink to the Soviet aesthetic references of the company.

Tag designed by Mishka NYC (front and back).

The front of the cardstock tags designed for the IVANAhelsinki collections shows the brand's logo, while on the back are hand-written the style, the color and the size of the garment.

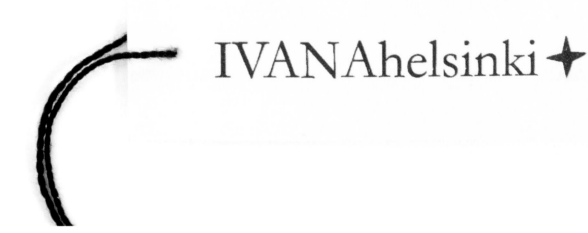

iVANAhelsinki | ivanahelsinki.com

HANDPRINTED

HANDMADE DETAILS

TRIKOOTUOTTEEN HOITO-OHJE:

- käsinpainettu tuote saattaa päästää väriä, erityisesti uutena tai joutuessaan suoraan hankaukseen
- pese nurinpäin käännettynä
- noudata pesuohjetta
- älä käytä valkaisevaa pesuainetta
- venytä tuote muotoonsa, kuivumaan asetellessasi
- vältä rumpukuivausta
- elastaania sisältävissä tuotteissa vältä huuhteluainetta
- huomioi kokoa valitessasi, trikoomateriaali saattaa kutistua

Tags designed by Paola Ivana Suhonen for the different IVANAhelsinki collections.

These tags for the Japanese restaurants specialized in sushi Itsu chain are, in fact, invitation cards for the local press on the occasion of the opening of the brand's establishments. Along with them come typical elements of the Japanese gastronomy such as chopsticks and tea bags.

why do the Japanese live so long...

Issue Number:

new brand, new concept, new store Vogue House, Hanover Square

itsu

1 chopstick tag
=
1 lunch on us

for more info, contact: *Afroditikrassa*
(t) +44 (0)20 76273463
press@afroditi.com

health
**because itsu food is light, full of
goodness & won't make you fat**

happiness
because it's not rabbit food

The point of itsu is low fat, fresh food NOT rabbit food. We hate
rabbit food. Leaves and seeds may be good for you, likewise
willpower and weightwatching. Easier said than done. Humans
crave toffee not tofu ... is it surprising?

Since we opened in 1997 we've battled to improve our dressings,
soups and salads. Light, healthy food can be delicious, it must
tingle ... it must make you feel good. We're making headway.

A Russian espionage drama put the spotlight on itsu the world
over. By a bizarre turn of fate this fame brought new people
to our doors.

Thanks to our customers and staff for their amazing
encouragement and resolve during this tricky time.

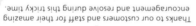

Invitation-tags designed by Afroditi Krassa for Itsu.

An old capsule of nitrous oxide (commonly known as laughing gas) has been turned into a key-chain, a present for the British clients who buy any Yes No Maybe product on the internet.

Capsule and tag designed by Ben Farleigh for Yes No Maybe.

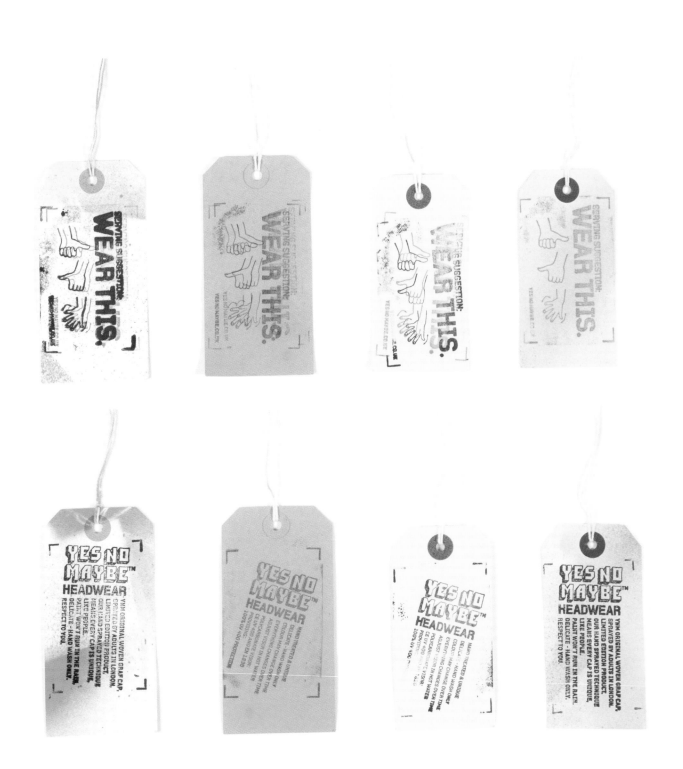

Complete line of tags designed by Ben Farleigh (front and back).

Printed on glossy cardstock and die-cut rounded corners, the tag for Yes No Maybe has not changed almost since the launch of the brand. The message on its back is not so much Yes No Maybe leit motif but rather one of its influences.

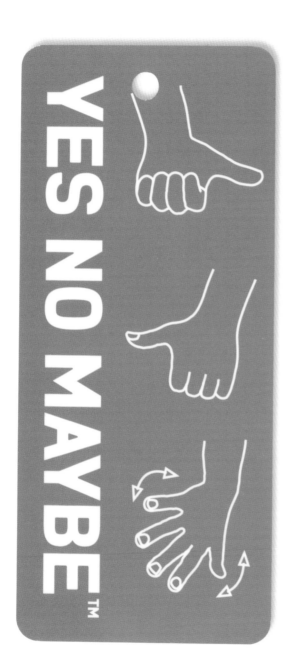

YES NO MAYBE™

The essence of existence is to make decisions. We are all bombarded with questions every day, trying to filter our constantly expanding choices. It's decision time, all the time. They teach us in school and the media that we can do anything if we want it enough and work hard enough.
So what do you want to do? YNM is for the generation that knows they don't know.

"The only true wisdom consists of knowing you know nothing"
- Socrates

yesnomaybe.co.uk

Tag designed by Brian Taylor (known as Candykiller) for Yes No Maybe.

Tags

The Dalliance font was chosen for the design of this tag, which is used in various Think Pink collections and follows the distinctive style of the brand. The retro style illustrations like those in an old naturalist's manual confer a romantic atmosphere.

Th!nk pink ♥

color.
talle.
precio.
art.

Tag designed by Mek Frinchaboy for Think Pink (front and back).

Psychedelic tag for the spring-summer 2008 collection of the fashion brand 47 Street. The designer hand-drew both the lettering and the illustrations.

Tag designed by Mek Frinchaboy for 47 Street (front and back).

Design by Mek Frinchaboy (front and back).

Art Nouveau was the inspiration and main influence for the design of the tags of 47 Street high-end line Premium collection. Printed in one color, the tags reflect the naïf air of the brand.

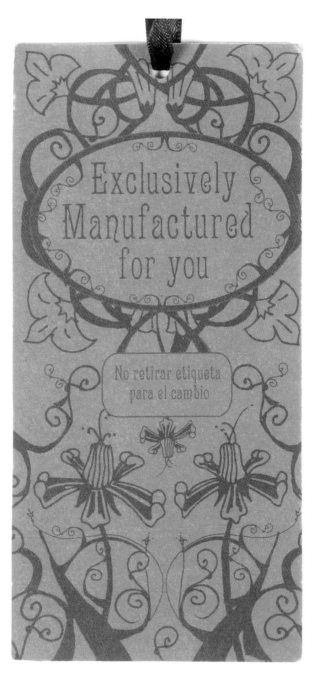

Tag designed by Mek Frinchaboy for 47 Street (front and back).

Design by Mek Frinchaboy (front and back).

Tag printed in two colors, black and fluorescent pink (the latter is also the color of the lace), for 47 Street line of summer dresses. The calligraphies have been hand-drawn, although in this case a photograph was added to the design.

Tag designed by Mek Frinchaboy for 47 Street (front and back).

Designs by Mek Frinchaboy.

The only requirement that the client had about the design of the tags and the graphic identity of 47 Street was that they would be "romantic" and "psychedelic." This explains the use of hand-written calligraphy and infantile graphic elements.

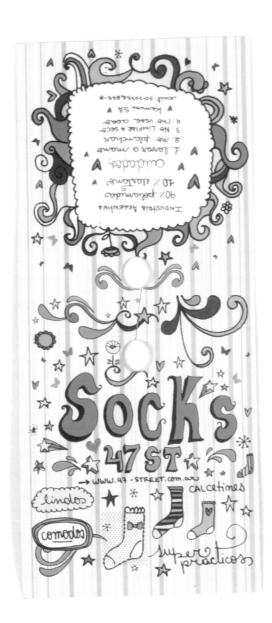

Folding kits for panty hoses made of cardstock. Design by Mek Frinchaboy for 47 Street.

The popular Japanese character Hello Kitty—created by Ikuko Shimizu for the company Sanrio in the seventies—is the protagonist of these tags for 47 Street. The red ribbon is similar to that worn by the character on its left ear.

Design by Mek Frinchaboy (front and back).

Embossed tags designed for 47 Street Rock collection. The font is Downcome, one of the few exceptions to the general use of the hand-written lettering that is typical on all the brand's tags.

girl power

47 STREET

cool wear

No retirar etiqueta para cambios

47 STREET

love!

www.47-street.com.ar

summer time

have enjoy

fun style

Cardboard tag designed by Mek Frinchaboy for 47 Street (front and back).

The tags for Lazy Oaf had to be fun, with a pop aesthetic. This explains the use of an animated character that symbolizes the brand's spirit and that, at the same time, brings essential information to the consumer.

hello,
these badges were designed by us at oaf towers. We like drawing and eating biscuits mainly. But If you like these then please take a look at our other lovely things on our web site. ♥
www.lazyoaf.com

lazy oaf © 2007/08

hello,
these things were designed by us at oaf towers. We like drawing and eating biscuits mainly. But If you like these then please take a look at our other lovely things on our web site. ♥
www.lazyoaf.com

lazy oaf © 2007-08

Tag design by Gemma Shiel for Lazy Oaf.

Design by Gemma Shiel, of Lazy Oaf (front and back).

Designed for Lazy Oaf's London store collections, the tags on this page show the brand's mascot in various colors and different expressions. The goal to humanize its appearance is achieved by presenting the information empathetically to the client.

Design by Gemma Shiel for Lazy Oaf.

Tag by Gemma Shiel for Lazy Oaf.

The promotional stickers for Lazy Oaf are sent to the press and are also included with the internet orders. The goal is to ensure the adolescent female clients' alliance to the brand by allowing them to "customize" the stickers and use them to decorate their accessories.

Promotional stickers designed by Gemma Shiel for Lazy Oaf.

Designs by Gemma Shiel for Lazy Oaf.

The tags for Romeo & Juliet Couture had to represent the brand without any added extemporaneous graphic elements. On this page, black is reserved for the brand's De Luxe line.

Tag designed by David Barath in collaboration with Visual Group for Romeo & Juliet Couture.

The tag designed for Paula Cahen d'Anvers' children line reinforces its marine references and invites the reading of a poetry that synthesizes the company's spirit, thanks to its book mark format.

Tags by Fbdi for Paula Cahen d'Anvers spring-summer 2008-2009 season children's collection.

Ketchup&Majo's logo was decorated with the silhouette of a tree made to look like a crown. This design was repeated on all the brand's graphic elements, from tags through to all types of promotional photographs.

Top: designs by Denis März and Janina Meyer of Ketchup&Majo for the Bread & Butter Barcelona 2006-2007 fair. / Bottom: tag designed by Denis März and Janina Meyer of Ketchup&Majo for the Schnarch recycling collection.

Red tag: design by Denis März and Janina Meyer of Ketchup&Majo for the recycling collection created for Topshop. / Black tag: design by Denis März and Janina Meyer of Ketchup&Majo for the winter 2007-2008 Alaaf collection.

Tags

Atelier LZC and Les touristes have designed a collection of suitcases and their tags. Square and simple, the tags play with various fonts and with the imbalance of the letters' base lines.

Design by Atelier LZC and Les touristes for Les touristes.

Design by Atelier LZC for Adieu Tristesse.

The Gold line of North Sails is the brand's most sophisticated and expensive. These tags stand out from all of the others because of the golden color of the fonts and embroideries that reach the string of the tag and the grommet.

Tags designed by Tomasoni Topsail for North Sails.

The Original is North Sails' most emblematic collection. The same materials have been used since the fifties (the first Lowell collection dated from 1958) and its tags, a dark blue, reflect the classical spirit of the brand.

Design by Tomasoni Topsail for North Sails.

The brand SKUNKFUNK often experiments with the different tag formats—in this case they're square and rectangular—that compliment with the garment or the line they adorn.

Design for unisex cardboard tags by Susana Sánchez Monje and Sergio Llanos of SKUNKFUNK.

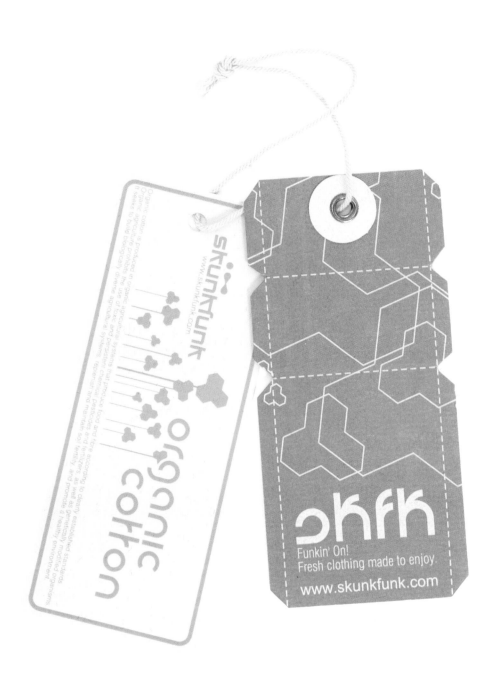

Tags

The double tag Organic of SKUNKFUNK reproduces the development of a box, including the flaps and goes with the garments fabricated with organic materials. The tag is printed on recycled cardstock and its rounded font was created especially for this collection.

Tags by Susana Sánchez Monje and Sergio Llanos of SKUNKFUNK.

Tags

Tags for SKUNKFUNK spring-summer 2008 season, men and women's collections. The tags for the women's line show rounded forms, while those for the men's line present abstract geometric forms and sharp angles.

Who cares about fashion pictures featuring perfect models? What we want to see you enjoying our garments. And that is what "People do Funk" is about. For all selected photos displayed on our web page we will send you in return an exclusive T-shirt of ours.

Send us your message together with your photo! Maybe you want to find your holiday love again who was wearing these cute knickers, or you would like to praise your mum's pasta sauce which has stained your favourite T-shirt.

Make an effort and you'll get in our "best of" gallery!

Design by Susana Sánchez Monje and Sergio Llanos of SKUNKFUNK.

Embroidered with green and red thread, the tags for SKUNKFUNK's Make Your Green fall-winter 2008-2009 collection make reference to the cold Spanish winter of 1938. A clear background and the silhouette of a leafless tree achieve the desired effect.

Double tags designed by Susana Sánchez Monje and Sergio Llanos of SKUNKFUNK.

On this page, tags printed on recycled cardstock belong to unisex collections fabricated with organic and ecological materials based on the "all green" concept. The fonts are corporate looking, while the graphics belong to the brand's different collections.

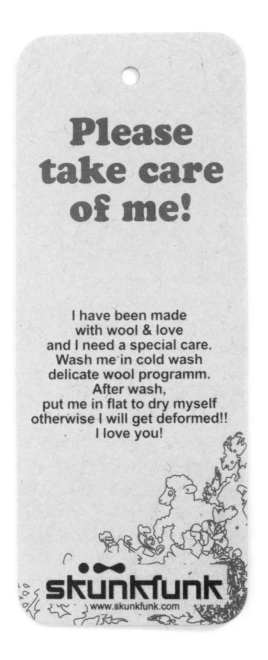

Please take care of me!

I have been made
with wool & love
and I need a special care.
Wash me in cold wash
delicate wool programm.
After wash,
put me in flat to dry myself
otherwise I will get deformed!!
I love you!

skunkfunk
www.skunkfunk.com

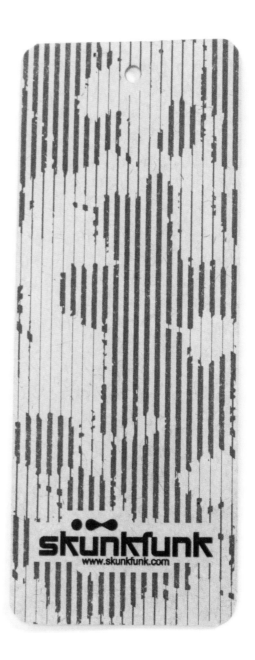

skunkfunk
www.skunkfunk.com

Designs by Susana Sánchez Monje and Sergio Llanos of SKUNKFUNK.

Round, decorated with an abstract graphic motif that makes reference to the Cuban sky, printed on cardstock and finished up with a metallic hanger, the tag for SKUNKFUNK's swimwear collection Waterfunk 2008 alludes to the brand's summery aesthetic.

Design by Susana Sánchez Monje and Sergio Llanos of SKUNKFUNK.

Other tricks SKUNKFUNK has used over the years for designing its tags have been to play with the die-cut of the tag (to the left) or use figurative illustrations like those of the fish (tag on the right).

Tags designed by Susana Sánchez Monje and Sergio Llanos of SKUNKFUNK.

Unisex tags for SKUNKFUNK's spring-summer 2009 collection. The decorative pattern represents a rainbow and other natural elements. Green is the brand's corporate color and recycled cardboard is the material used.

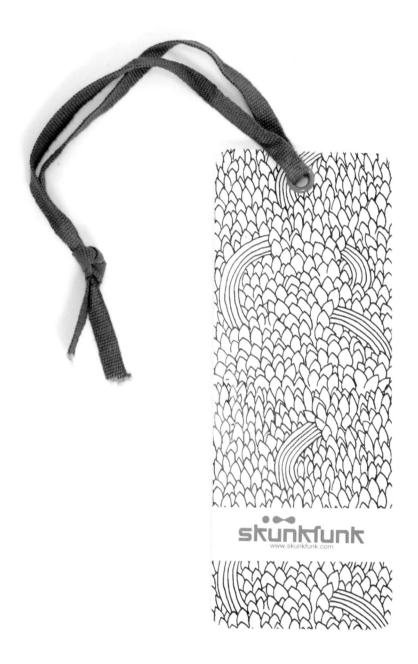

Design by Susana Sánchez Monje and Sergio Llanos of SKUNKFUNK.

innocent®
pure fruit smoothie
blackberries & blueberries

innocent®
pure fruit smoothie
cranberries & raspberries

innocent®
pure fruit smoothie
mangoes & passion fruits

Labels

It would not be an exaggeration to say that a simple label carries out more simultaneous functions than an air traffic controller in the midst of a snow storm: it informs about the brand and provides contact information, instructs on the correct use of the product and on its technical characteristics, offers useful information regarding its origin and conveys all the aesthetics and even "vital" values of the brand. Also, unlike air traffic controllers, it has to be graphically harmonious. Is it a possible task?

5,0 Original is a low cost brand for the German market. The simple label, for which only two colors have been used, is reflected on the beer box, a minimalistic design that strives to capture the client's attention without any added detail.

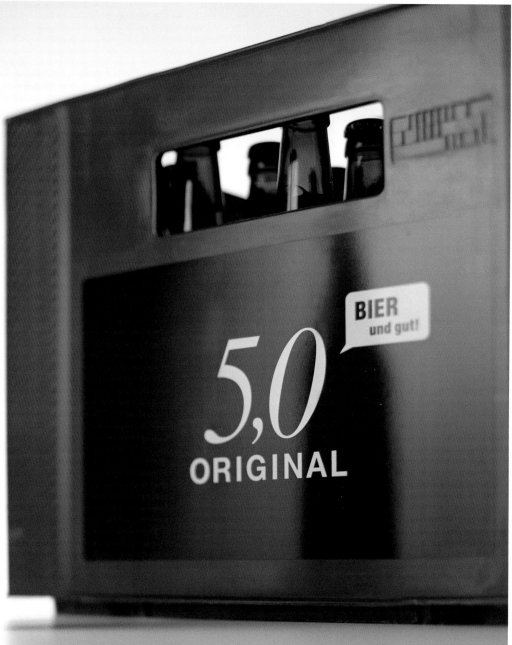

Design by Feldmann+Schultchen Design Studios for Carlsberg Deutschland Gruppe's 5,0 Original standard bottle.

Line of labels and color codes by Feldmann+Schultchen Design Studios.

The line of labels 5,0 Original uses the traditional color codes in the German beer market: white and black for Pilsner, red and white for Export, etcetera. The fonts are Helvetica and Times.

Line of cans designed by Feldmann+Schultchen Design Studios for Carlsberg Deutschland Gruppe.

Design by Pablo Martín and Bárbara Castro of grafica for Camper.

Composed of three types of paper that are reminiscent of the way the traditional kimono fabrics superimpose over each other, the label created for the Kivino project was presented at the Design Escapade Show in Turin in 2008.

Design by Setsu & Shinobu Ito for Mylena.

Fabricated for the Spanish Cava Regulatory Council, the tags designed by grafica make sense when placed one next to the other: each of the four tags has been decorated with one of the letters of the word "cava."

Labels designed by Pablo Martín and Ellen Diedrich of grafica
for the Regulatory Council of Cava.

When Steinlager Pure stopped being considered "the best beer in New Zealand," it was time for the product and its design to be renewed. To achieve this, Trajan and Helvetica fonts, as well as transparent labels, were used to emphasize the product's quality.

Design by Publicis Mojo and Lynne Richardson of Curious Design Consultants for Steinlager Pure.

Digitally re-touched image of the Steinlager bottle Pure
for its presentation at the selling points.

The label designed for Sero2 reflects the characteristics of the product by means of a harmonic image and the interaction of Bauhaus Standard and Santana fonts. The simplicity of the design attracts the attention of the consumer by contrasting with the competition's more florid designs.

serotonin naturally

SERO²

Softly Sparkling
Serotonic
Spring Water

Label design by Natalie Dawson of Curious Design Consultants
for New Zealand Aquaceuticals.

Using the fonts Trajan and Futura, and the colors PMS 877, Siver Foil and, PMS 363, Curious Design emphasized the New Zealand origin of Christie's gin. The use of the transparent tag reinforces the purity of the liquor.

ESSENCE *of* AOTEAROA

QUADRUPLE DISTILLED AND BLENDED WITH AROMATIC BOTANICALS, SUBTLY FINISHED WITH A HINT OF NATIVE HOROPITO & KAWAKAWA

Christie's

Christie's

CRISP

NEW ZEALAND

GIN

40% ALC/VOL 1 Litre

Design by Damian Alexander of Curious Design Consultants for New Zealand gin Christie's.

For the design of Saint Clair's complete graphic image a faded and irregularly die-cut label was created to convey the impression of antiquity. The fonts (Matrix Bold, Shelly Allegro Script and Rosewood Fill) reinforce this impression and appear as if hand-written.

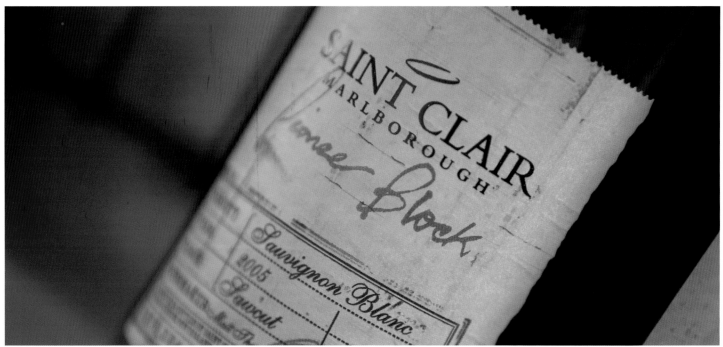

Design by Tony Ibbotson, Andy Yanto and Mayra Minobe
of The Creative Method for Saint Clair Family Estate.

Line of labels by The Creative Method for Saint Clair Family Estate.

For the retro style redesign of the packaging for the Real McCoy's liquors, emphasis was put on the history of the brand, on the elements that differentiate its products from those of the competition and on the details that make their liquors a high-end product. The goal is to make its image stand out amongst those of its competitors.

Label design by Tony Ibbotson, Andy Yanto and Mayra Minobe
of The Creative Method for Diageo Australia.

The labels for Guzman y Gomez's spicy sauces show the image of the brand's store managers with expressions that identify very graphically how spicy each of the sauces are, from "medium" to "extra spicy." The label includes a comment from the managers on this particular sauce.

Design by Tony Ibbotson, Andy Yanto and Mayra Minobe
of The Creative Method for Guzman y Gomez.

Design for the line of labels and disposable glasses by The Creative Method.

Elbows Bend is a small wine company in New Zealand that uses only grapes from of the best regions in the making of its wines. This explains why The Creative Method, who designed the label, bet on simplicity and classical fonts (Bodoni Book, Univers Light).

Design by Tony Ibbotson, Andy Yanto and Mayra Minobe of The Creative Method.

Design for a line of labels by The Creative Method.

Grumblebone is a small store in the Australian Southern Highlands specialized in hand-crafted wines. The simplicity of the labels contrasts with the humoristic illustration of a grouchy character who decorates the bottle's neck and with whom the clients associate the brand.

Design by Tony Ibbotson, Andy Yanto and Mayra Minobe of The Creative Method for Paul Sibra.

The dye-cut tag on this page has been designed with a scanned image of a scorch mark on a paper that crosses the brand's logo. The different colors of the tags correspond to the diverse varieties. The fonts used are Futura Condensed, Galliard, and Bodoni.

Labels design by The Creative Method for Marlborough Valley Wines.

In Catalonia (Spain), the *pubilla* is the eldest daughter in a family. This sparkling wine (cava) has been created to honor the first daughter of Pere Ventura, who is the founder of the winery that carries his name. The design of the label received the Red Dot Design Award and the Gold medal at the Pentawards in the Luxury category.

White glossy propylene label. Matt black, red, silver silk-screen printing, glossy
varnish, and gold stamping designed by José Mª Morera of Morera Design
for Pere Ventura's cava La Pubilla.

To unify the composition of a metallic label and glass bottle for the Miller Chill beer, a green color was selected, which was also a subtle wink to the Mexican origins of the drink's formula (green is often associated with Mexican gastronomy). The font is Bickham Script along with other hand-written lettering.

Label designed by Adam Ferguson (designer) and Lyle Zimmerman (art director)
of Optima Soulsight for Miller Brewing Company.

The tag on the Smirnoff bottle on this page reflects the main characteristics of the vodka: It is suave, refreshing, and containing an abundance of minerals. This explains the use of white and the transparencies.

Design by London Tokura of SiO DESIGN for Diageo Japan KK.

For Absolut 100, a minimalistic and masculine pal-
ette over semi-opaque black glass was chosen to
contrast with a chromed silver type and grey text.
A transparent logo that suggests sensuality and
luxury was selected for Absolut Vanilia.

Designs by Spedding Westrip and Lisa Simpsom of Pearlfisher for Absolut 100
and Absolut Vanilia, respectively.

NEW ORLEANS

40% ALC./VOL. (80 PROOF) 750 ML.
IMPORTED
PEACH FLAVORED VODKA
PRODUCED AND BOTTLED IN ÅHUS, SWEDEN.
V&S VIN&SPRIT AB (PUBL)

ABSOLUT
CUT.

UNIQUE SWEDISH SPIRIT TONIC WITH
Imported Absolut Vodka
A FRESH SPARKLING CITRUS TASTE
7% alc./vol. 330 ml

ABSOLUT®
Country of Sweden
MANDRIN

Absolut Mandrin is made from
a unique blend of natural mandarin
and orange flavors and vodka
distilled from grain grown in
the rich fields of southern Sweden.
The distilling and flavoring of vodka
is an age-old Swedish tradition.
Vodka has been sold under the name
Absolut since 1879.

40% ALC./VOL. (80 PROOF) 750 ML.
IMPORTED
MANDARIN FLAVORED VODKA
PRODUCED AND BOTTLED IN ÅHUS, SWEDEN
BY THE ABSOLUT COMPANY
A DIVISION OF V&S VIN&SPRIT AB

ABSOLUT
Country of Sweden
RASPBERRI

Feel the intense burst of
natural Raspberry, blended with
vodka distilled from grain grown
in the rich fields of southern Sweden.
The distilling and flavoring of vodka
is an age-old Swedish tradition,
dating back more than 400 years.
Vodka has been sold under the name
Absolut since 1879.

40% ALC./VOL. (80 PROOF) 750 ML.
IMPORTED
RASPBERRY FLAVORED VODKA
V&S VIN&SPRIT AB (PUBL)

level.

IMPORTED
VODKA
Spirit of Absolut
40% ALC./VOL. (80 PROOF) 1 LITER

Designs by Pearlfisher for The Absolut Spirits Company.

For Tennent's strategic repositioning, Pearlfisher decided to emphasize the elements that suggest excellence and quality, and pay homage to its Scottish origins (Glasgow) with a brief and proud informative text.

Design by Sarah Carr of Pearlfisher for InBev UK.

Line of labels designed by Lisa Simpson of Pearlfisher for the wine-producer (oops).

This Water is a new brand of spring water created by Innocent Drinks. The complete line is composed of four flavors, and each of them is provided with two different design labels (eight labels total). If the line of flavors expands, the flexible design will adapt without difficulty.

this water™
is made from
fruit and clouds

lemon, lime & spring water

cranberries
for health,
this water™
for thirst

cranberries, raspberries & spring water

this water™
is coloured by nature

mango, passion fruit & spring water

42 blackcurrants
from Herefordshire
found a good home in
this water™

blackcurrants & spring water

fruit from
the trees,

this water™

from a spring

mango, passion fruit & spring water

this water™

quenches thirsts
for a living

blackcurrants & spring water

no trucks came from
the Alps to bring you

this water™

lemon, lime & spring water

there are 89
cranberries in
every bottle of

this water™

and 13 raspberries too

cranberries, raspberries & spring water

Design by Sarah Pidgeon of Pearlfisher for Innocent Drinks.

Woolworths is a well-known and prestigious South African retailer with no relation with Woolworths Group. Pearlfisher has designed and simplified the brand image of its more than 4,000 products, unifying its aesthetics and adapting it to the new trends.

Designs by Natalie Chung of Pearlfisher for Woolworths.

Txin! Txin! is, other than the name of this sparkling wine (cava), the expression used in Catalonia for toasting. The minimalist black label reinforces the social and the communal meaning of drinking, and provides the product with a subtle anthropomorphic character.

Label designed by Pablo Martín and Borja Martínez of grafica for the Auditori de Sant Cugat.

Maltana is an alcohol-free malt drink available in four different flavors, targeted at the young person who does not drink alcohol but wants to seem "cool" at parties. This resulted in a tag that follows the usual masculine graphics of the conventional alcoholic drinks such as beer.

Design by David Rault and Burkan Ciftciguzelli of Paristanbul
for the alcohol-free malt drink Maltana by Ülker.

The lines that traditionally indicate the score in the games are used for the design of Project 7's products' brand image. So much so that the client visualizes and quickly identifies the products' seven humanitarian goals.

Label and box design by Tyler Merrick, Darren Dunham and Jonathan Rollins of 29 Agency for Project 7.

Complete line of labels. Design by Tyler Merrick, Darren Dunham
and Jonathan Rollins of 29 Agency for Project 7.

The recycled paper bag that comes with the product (high quality olive oil) prevents staining the client's hands, while associating it to ecological products. If one prefers to take the bottle out of the bag, the label remains stuck to the container.

Design for the label and the recycled paper bag by Davor Bruketa, Nikola Zinic and Ruth Hoffmann of Bruketa&Zinic for Chiavalon's Ex Albis.

The Boskinac brand includes three different lines of business—a winery, a restaurant, and a hotel—all located on a Croatian island. Its visual identity reflects the three references of its native island: the sun, the sea, and the barren land, thus the design of three different labels.

Labels by Davor Bruketa and Nikola Zinic of Bruketa&Zinic for the Croatian wine producer brand Boskinac.

The design uses a bee, starting with the shape of a honey spoon, as the image for Deluxe Honeydrop. The bee is a friendly and flexible icon permitting color combinations that identify the different honey flavors.

Design by Lisa Simpson of Pearlfisher for Deluxe Honeydrop.

This line of shower gel made by Waitrose has unique personality—each type has a different photograph, which acts as the visual representation of the aroma of the gel, along with clean and simple typography.

Complete label line designed by Pearlfisher for the shower gel collection Waitrose.

The label for the Summer Draft beer needed to be adapted to the original and asymmetrical shape of the bottle. This lead to strictly geometric fonts and to the transparent label that allows the bottle's silhouette to be the strongest visual element.

Label designed by Dil Brands for Summer Draft.

Line of labels designed by Dil Brands for Diet V8 Splash.

The folding aluminum label for the 2 cl BACARDI Corto glasses is based on the idea of "instant seduction." BACARDI Corto, aimed to the high market segment, shows the traditional bat, which identifies the brand's engraving on the glass.

Labels designed by Feldmann+Schultchen Design Studios for BACARDI Corto.

Worthy of a Silver Pentaward in 2008, the brand's image design for Liquid Series of Doutor Coffee uses the real colors of the drinks in their containers. The result both varied and homogenous from a graphic design point of view.

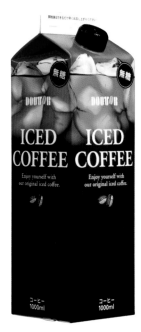

Designs by Akihiro Nishizawa of EIGHT for Doutor Coffee.

The boxes for the products of the Liquid Series line and the containers in them are a different color so that distributors and vendors can distinguish them. A simple design that categorizes the product within the segment of high-end products.

Designs by Akihiro Nishizawa of EIGHT for Doutor Coffee.

The premium line of coffee beans for Doutor Coffee are selected in alternate months. The container's design is always based on the history of this particular type of bean and uses different colors for each of the seasons to capture the attention of the client.

Designs by Akihiro Nishizawa of EIGHT for Doutor Coffee.

The design for the Premium Beans Selection line is sufficiently flexible so it can be adapted to the different types of containers of the line. The disposition of the beans that the front of the label illustrates changes based on the variety it identifies.

Designs by Akihiro Nishizawa of EIGHT for Doutor Coffee.

The design for the Premium Beans Selection line received a Pentaward Blond Award in 2008. The product's promotional posters use the particular disposition of the beans on the label as a distinctive image and is easily identifiable by the consumer.

Designs by Akihiro Nishizawa of EIGHT for Doutor Coffee.

Using a unique design for the packaging of a product and the various promotional elements allows for a strong brand image, especially if both are presented together at the selling points. Prices are always shown in the same place on the label to be quickly identified.

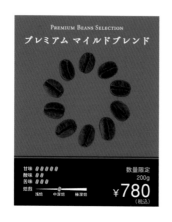

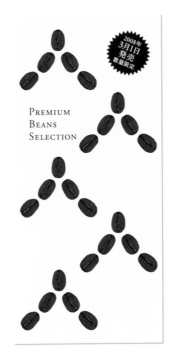

Designs by Akihiro Nishizawa of EIGHT for Doutor Coffee.

Doutor Coffee produces six different series of the Premium line every month. Each of them identified with their own particular design and an identifiable color. The final design goal is to reinforce the originality of the line's flavors and aromas.

PREMIUM BEANS SELECTION

Premium Brazil Fox Beans

数量限定
200g
¥1,070
（税込）

きつね色に輝く奇跡の逸品
プレミアム ブラジル フォックスビーンズ

完熟した実に多く含まれる糖質が、豆表面の薄皮を
きつね色に染める…。それは自然が生み出した特別な豆の証。
色彩選別機にかけられた無数の豆の中から、一粒ずつ
ハンドピックされた逸品は、シルクのような口当たりに、
ワインの香りとブラウンシュガーのような甘味をもつ、
上品な味わいに仕上がりました。

PREMIUM BEANS SELECTION

プレミアム ブラジル
フォックスビーンズ

甘味
酸味
苦味
焙煎　浅煎　中深煎　極深煎

数量限定
200g

¥1,070
（税込）

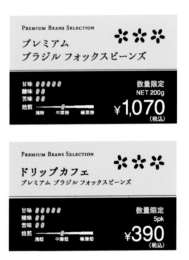

PREMIUM BEANS SELECTION

プレミアム
ブラジル フォックスビーンズ

甘味
酸味
苦味
焙煎　浅煎　中深煎　極深煎

数量限定
NET 200g

¥1,070
（税込）

PREMIUM BEANS SELECTION

ドリップカフェ
プレミアム ブラジル フォックスビーンズ

甘味
酸味
苦味
焙煎　浅煎　中深煎　極深煎

数量限定
5pk

¥390
（税込）

Designs by Akihiro Nishizawa of EIGHT for Doutor Coffee.

Pearlfisher's design for the food to go line Waitrose believes, almost exclusively, in the potential of the product's photographs and in the evocative power of a single word. A simple typography and evocative colors finish up the design.

Designs by Kate Marlow of Pearlfisher for Waitrose.

The design of the labels and the black bottles (to protect the product from the light) for the Italian line of olive oil; Waitrose has taken into account the different origin of each of the varieties. The typography has been selected for its similarity with that used on the olive producers denominations.

Designs by Natalie Chung of Pearlfisher for Waitrose.

peach kernel & vanilla
exfoliating body wash
for glowing skin

umi.

white rice & wheatgerm
bath soak to relax
mind & body

umi.

lotus flower &
oat ceramide
face cream
for radiant skin

umi.

vanilla & cardamom
body soufflé
for velvety skin

umi.

vanilla & brown sugar
bath grains to
soothe mind & body

umi.

Designs by Pearlfisher for the line of beauty and bath products Umi.

With the objective to create the first Turkish high-end pack of chewing gum for the country's growing middle class, Paristanbul chose a minimalist design: Simple typographies (Gill Sans), one single color per flavor and three different types of containers: cardboard, plastic and metal.

Designs by David Rault of Paristanbul for Ülker's Colors line of Chewy gums.

The new design for Union Hand-Roasted uses a simple and direct symbolism influenced by the company's conception as a union of farmers and entrepreneurs. Each variety of coffee is identified with one single symbol and a particular color.

UNION®
HAND-ROASTED
brazil
A subtle, sweet coffee whose gently ELEGANT flavour hints of cocoa and almonds can be TRUSTED by even the most fragile morning people
2: gentle

elegant
TRUST

UNION®
HAND-ROASTED
guatemala
The DEEPLY complex coffee with smoky butterscotch notes, whose lingering finish should be enjoyed after dinner with a PATIENT dining companion.
4: medium /strong

deep
PATIENCE

UNION®
HAND-ROASTED
colombia
The ALL DAY, all night coffee with smooth chocolate and fruity red wine notes for easy, HONEST drinking
3: medium

all day
HONESTY

UNION®
HAND-ROASTED
bright note™
The SMOOTH all day long coffee with a naturally sweet character, whose hazelnut and almond overtones hit the spot PRECISELY
3: medium

smooth PRECISION

UNION®
HAND-ROASTED
decaf
BOLD, sweet and indulgent, this coffee's rich caramel tones and smoky aroma will expand your UNDERSTANDING of decaf
4: medium /strong

bold
UNDERSTANDING

UNION®
HAND-ROASTED
Exclusive to Sainsbury's
sumatra
The DARK indulgent coffee with herbal chocolate aromas. For an INSIGHTFUL afternoon, enjoy with an equally peaty single malt
4: medium /strong

dark INSIGHT

Designs by Sean Thomas of Pearlfisher for Union Hand-Roasted.

This limited edition drink has been conceived as an answer to "the need for women's beauty." The design strengthens the image of freshness, naturality and luminosity and makes and refers to the different seasons.

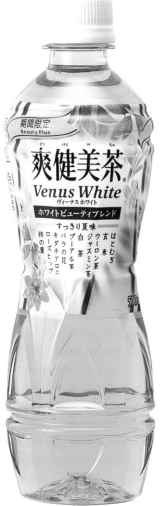

Design by London Tokura of SiO DESIGN for Sokenbicha Beauty Plus
of The Coca-Cola (Japan) Company.

The design of unusual simplicity and purity for Americano Black coffee conveys a brand's image that strengthens the idea of "blackness" and flavor "power" in a market dominated by products much more suave such as "café au lait."

Design by Yoshinori Shiozawa of SiO DESIGN for Americano Black
of UCC Ueshima Coffee.

Top: design by Kyoko Ebisawa of SiO DESIGN for Barista's Stile of UCC Ueshima Coffee. / Bottom left: design by Yoshinori Shiozawa of SiO DESIGN for Café Bruno of UCC Ueshima Coffee. / Bottom right: design by Yoshinori Shiozawa of SiO DESIGN for Kuro-no-Masamune of UCC Ueshima Coffee.

The design of this bottle evokes the purity and the "natural" origin traditionally associated to mineral water. The graphic pattern makes reference to the melting snow on the peaks of the Japanese Hida Mountains (also known as "Alps of the North").

Design by Yoshinori Shiozawa of SiO DESIGN for Azumino Yusui of Tokyo Art.

Design by London Tokura of SiO DESIGN for carbonated water brand Dasani of The Coca-Cola (Japan) Company.

Strength (or flavor intensity) and status (or exclusivity). These are the two characteristics to which the consumer immediately associates the high-end tea line contained in this aluminum tin that stands out for its retro design.

Design by Kyoko Ebisawa of SiO DESIGN for Black Tea of UCC Ueshima Coffee.

Metallic colors and warm tones on high end products like those on this page are associated with excellence. These color combinations are often used by designers when targeting opportunities for new, "exclusive" markets.

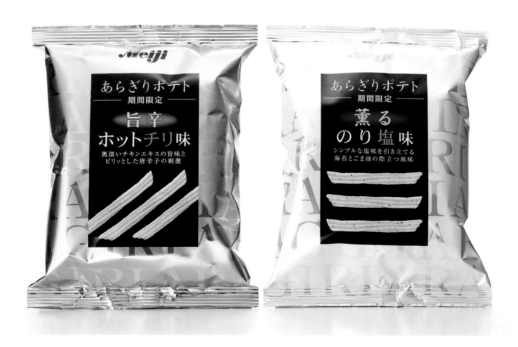

Top left: design by Yoshinori Shiozawa of SiO DESIGN for Cocoa No Zeitaku of Seika Kaisha. / Top right: design by Yoshinori Shiozawa of SiO DESIGN for Pure Cocoa of Meiji Seika Kaisha. / Bottom: design by Yoshinori Shiozawa of SiO DESIGN for Aragiri Ptato with Salt Aroma & Hot Chili Flavor of Meiji Seika Kaisha.

The design by Pearlfisher for the new Coca-Cola Blak with coffee flavor reinforces the association of the product with the characteristics of "intense flavor" and "mysticism." The selected icon is an abstract coffee bean created from Coca-Cola's traditional ribbon on the logotype.

Design by Sean Thomas of Pearlfisher for The Coca-Cola Company.

The redesign of the tags for the Dr. Stuart's tea brand presents the company as an expert in all kinds of herbs. This explains the use of surrealistic illustrations that make reference to the diverse varieties of tea, the use of white as a background, and the peculiar but very representative descriptions.

Designs by Natalie Chung of Pearlfisher for Dr Stuart's tea line.

The love for gardening was the starting point for the design of the line of Fortnum & Mason's hand-crafted jams and preserves. In fact, the label adopts the form of the traditional little signs for plants used in traditional gardening.

Designs by Sarah Carr of Pearlfisher for Fortnum & Mason.

The tags for the Fortnum & Mason tea line are elegant and slightly retro. The objective is to avoid breaking the harmony and the sweet, romantic aesthetic of the product's embossed metallic boxes.

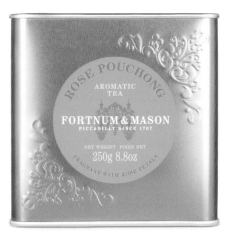

Label design by Pearlfisher for the tea line of Fortnum & Mason.

For the redesign of the graphic identity of Innocent Drinks, Pearlfisher touched up the halo of the image so that "it would not resemble a baguette." The character reinforced its proud personality as the brand's icon.

Designs by Pearlfisher for Innocent Drinks.

Jme is the brand of 170 products for the home and the garden created by the famous British cook Jamie Oliver. Pearlfisher was commissioned to design the graphic image of the brand. Each product needed a graphic solution to make it quickly identifiable.

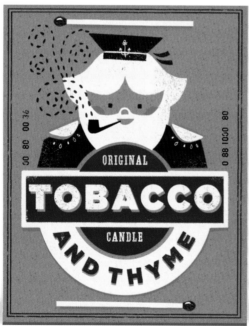

CHILLI SEA SALT

a pinch of
ROSE MARY
sea salt

she sells sea
salt on the sea
shore. I'm sure
it's Citrus
sea salt
for sure

Designs by Sarah Pidgeon and Natalie Chung of Pearlfisher
for the Jme line of Jamie Oliver.

The logotype for Jme visually unifies the eclectic line of graphic designs created for each product. In most cases, all the wrappers and containers are recyclable and reusable.

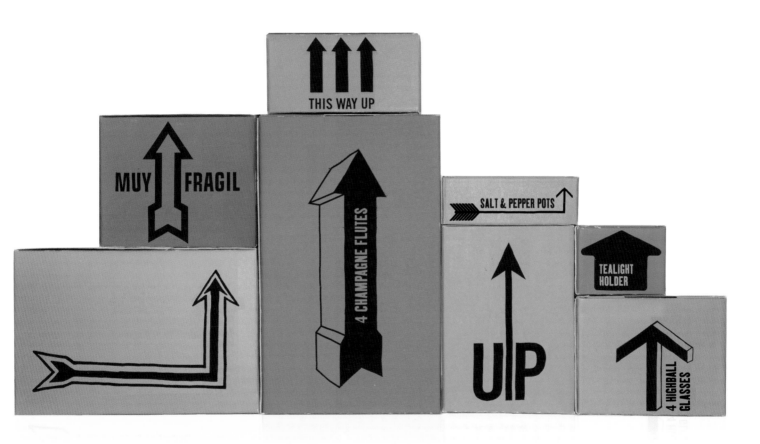

Designs by Sarah Pidgeon and Natalie Chung of Pearlfisher
for the Jme line of Jamie Oliver.

Labels

Jme products relate to the consumer at an emotional level thanks to the slogans and the personalized style of the informative texts included on the labels. The brand's personality is perceived as "fresh and relaxed."

Designs by Sarah Pidgeon and Natalie Chung of Pearlfisher
for the Jme line of Jamie Oliver.

The illustrations, the logos and the informative texts designed for Jordans decorate all four sides of the container. The color of each of the cereal varieties contrasts with the white background for the brand's logo.

Designs by Will Gladden of Pearlfisher for Jordans.

For the brand's image redesign for Òutspan drinks, a new typography and photographs of fruit were used. The halo on the initial of Òutspan is intended to reflect the natural origin of the ingredients used in the juice.

SMR Creative for Òutspan of Aimia Foods.

Designs by SMR Creative for Slazenger.

The design of the labels for Azienda Agricola Rigoc-cioli uses geometric forms, softened by hand-drawn illustrations that show the ingredients of the food. Four colors were chosen: one for each type of product (gelatins, preserves, oil preserves and patés).

Gaia Bisconti Design for Azienda Agricola Rigoccioli.

The inverted nutritional pyramid and the slogan "pyramid of the new school, nutrition of the old school" make reference to an era in which food for pets used high quality ingredients. The containers for dog food have a black background while those for cat food have a white background.

Tyler Merrick, Darren Dunham and Jonathan Rollins of 29 Agency for Merrick Pet Care.

Modern labels, simple and neutral, designed for the tea bags at Foodelicious deviate all the attention towards the product, which has to compete with other similar products with much more colorful labels and packaging. The font used is ITC Avant Garde Gothic.

Labels designed by Marjolein Delhaas for Foodelicious.

Label and container designed by Afroditi Krassa for the British chain sushi restaurants Itsu.

Labels

The closing label shown on this page was designed to be used on all the bento boxes for sushi. The models are the traditional Japanese vertical bands. The white background allows the logo to be seen clearly. The fonts used are Flux and Frutiger.

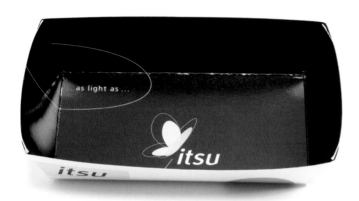

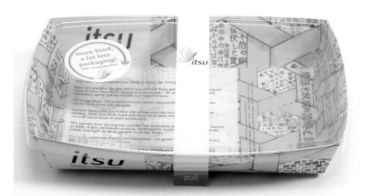

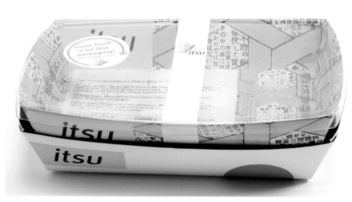

Yoghurt Fruit Goji Berries

Ingredients

Banana Chips, Goji Berries, Pineapple, Raisins, Vegetable Fat, Sugar, Yoghurt Powder, Skim Milk Powder, Whey Powder, Glazing Agent, Emulsifier, Cranberries, Blueberries.

Allergen Information

These products are manufactured in a kitchen where wheat, gluten and sesame seeds are processed. Avoid if allergic to nuts.

Nutritional Information - per pack

energy	320.1 kcal
protein	30.8 g
fat	11.2 g

70g ℮   Best before: see pack for details. Packed in the UK for itsu Ltd. www.itsu.com

itsu
health & happiness

Closing label and container designed by Afroditi Krassa for Itsu.

The abstract graphics used to decorate these labels for Itsu are, actually, close-ups of old Japanese postcards of the 16th century. Bright colors such as Pantone Rhodamine Red and Pantone Yellow 108 are used on the colored boxes.

Peanut Rice Crackers

Peanut Rice Crackers

Ingredients

Rice, Shoyu Sauce, Peanuts, Chilli Pepper, Seaweed, Natural Food Colours: Capsanthin (Red) E160C, Annatto (Red) E160B, Annatto (Yellow) E160B, Chalorophy (Green) E140.

Allergen Information

These products are manufactured in a kitchen where wheat, gluten and sesame seeds are processed. Contains nuts

Nutritional Information - per pack	
energy	264 kcal
carbohydrates	55.8 g
protein	6.3 g
fat	3.3 g

70g ℮   Best before: see pack for details. Packed in the UK for itsu Ltd. www.itsu.com

itsu
health happiness

Cardstock labels designed by Afroditi Krassa for Itsu.

Labels

The labels designed for Itsu's glasses and bowls used for hot food show the already mentioned traditional Japanese patterns of the 16th century, with the only exception being the bowl for miso soup which is left white to distinguish it from all others.

Labels for disposable glasses and bowls designed by Afroditi Krassa for Itsu.

itsu

Vegetarian Gyoza,
Asian Vegetables,
Udon Noodles,
Homemade
Miso Broth

349 kcal   1.8% fat

itsu

Homemade
Miso Broth,
Wakame,
Tofu,
Coriander

19 kcal   0.5% fat

only
15 kcal
0.5% fat

**Skinny
Miso**

Skinny Facts

A delicious combination
packed with nutrients,
vitamins and minerals.

Made with wakame, fresh
vegetables and silken tofu.

Enjoy with a salad, a bento
box or on its own.
Morning, noon or night.

The label-watch (top left) permits the workers at Itsu to indicate the expiring date of the served dish, which emphasizes the commitment with the food's freshness of the British brand. Other labels show information on the product.

the nutritional benefits of our low calorie, low fat, vitamin fuelled hot dishes are outstanding.

miso is admired for its antioxidant & immune boosting properties. nutritionists & dieticians claim it's an outstanding source of protein, even lowering cholesterol.

eat with a fork, spoon, chopsticks or simply slurp

more food, a lot less packaging! (than our previous box)

Labels for glasses and bowls designed by Afroditi Krassa for Itsu.

Labels

The label for Axe's shower sponge is designed using two fonts: Universe Black Extended and Inter-Face DaMa Bold. The metallic label allows the string on the product to be reflected on it so that the client can see it quickly.

Graphic design by JOED Design for Unilever. Product design by Prime Studio.

A simple design of two warm tones in the brown range allows the client to associate the product with its main ingredient. The darkest tones suggest a higher concentration level of cocoa.

Design by Kyoko Ebisawa of SiO DESIGN for Meiji Seika Kaisha.

A transparent tag decorated with an orange dragon, which makes reference to the geographical origin of the product (green tea from Far East), remains a background element and allows the client to focus on the design of the product's spectacular packaging.

Green Tea Gel Beads by swerve

Label and container design for Swerve.

In this case, the label and the packaging of the product have been designed by two creative teams, which required a great deal of coordination. The colors and the fonts, for instance, are those used in the graphic image of Boost Mobile, preventing radical innovations.

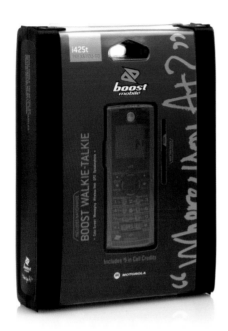

Design of packaging by Swerve for Boost Mobile. Label designed
by Boost Mobile.

The labels designed by Marjolein Delhaas for Food-elicious line of products stand out for their formal simplicity and for their subtle selection of colors. This allows differentiating the products and playing with the design to adapt it to the different seasons, adding gold color for Christmas for instance.

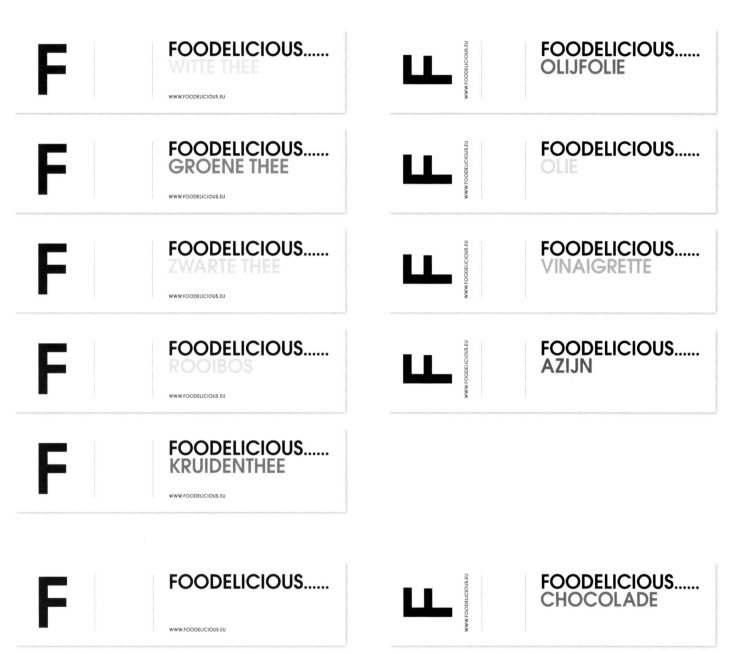

F FOODELICIOUS......
WITTE THEE
WWW.FOODELICIOUS.EU

F FOODELICIOUS......
GROENE THEE
WWW.FOODELICIOUS.EU

F FOODELICIOUS......
ZWARTE THEE
WWW.FOODELICIOUS.EU

F FOODELICIOUS......
ROOIBOS
WWW.FOODELICIOUS.EU

F FOODELICIOUS......
KRUIDENTHEE
WWW.FOODELICIOUS.EU

F FOODELICIOUS......
WWW.FOODELICIOUS.EU

LF FOODELICIOUS......
OLIJFOLIE
WWW.FOODELICIOUS.EU

LF FOODELICIOUS......
OLIE
WWW.FOODELICIOUS.EU

LF FOODELICIOUS......
VINAIGRETTE
WWW.FOODELICIOUS.EU

LF FOODELICIOUS......
AZIJN
WWW.FOODELICIOUS.EU

LF FOODELICIOUS......
CHOCOLADE
WWW.FOODELICIOUS.EU

Designs by Marjolein Delhaas for Foodelicious.

The labels for Foodelicious give predominance to the name of the product and do without any other type of added information, except for the name of the brand and, in some cases, the web page. An image, fresh and free of artifice, is then created.

WWW.FOODELICIOUS.EU

WWW.FOODELICIOUS.EU

FOODELICIOUS..... **THEE**

FOODELICIOUS..... **VINAIGRETTE**

FOODELICIOUS..... **OLIJFOLIE**

FOODELICIOUS..... **AZIJN**

WWW.FOODELICIOUS.EU

WWW.FOODELICIOUS.EU

Designs by Marjolein Delhaas for Foodelicious.

The original plastic boxes by contemporary jewelry designer Charlotte Wooning show around the perimeter a tag with the names of the 12 stones (Helvetica Neue Bold) used in the necklaces. Each stone possess alleged properties. Just a slight color touch indicates the type of stone a particular piece of jewelry is made of.

aventurine make me adventurous tourmaline make me pretty & sexy agate make me good, better, best amethyst make me clairvoyant malachite make me positive & passionate lazuli make me love & loved moonstone make me feel feminine & mysterious amber make me as good as new jade make me confident & happy pearl make me weep & forget citrine make me warm & tender chalcedony make me dream & fantasize chainchains. charlottewooning

aventurine make me adventurous tourmaline make me pretty & sexy agate make me good, better, best amethyst make me clairvoyant malachite make me positive & passionate lazuli make me love & loved moonstone make me feel feminine & mysterious amber make me as good as new jade make me confident & happy pearl make me weep & forget citrine make me warm & tender chalcedony make me dream & fantasize chainchains. charlottewooning

aventurine make me adventurous tourmaline make me pretty & sexy agate make me good, better, best amethyst make me clairvoyant malachite make me positive & passionate lazuli make me love & loved moonstone make me feel feminine & mysterious amber make me as good as new jade make me confident & happy pearl make me weep & forget citrine make me warm & tender chalcedony make me dream & fantasize chainchains. charlottewooning

aventurine make me adventurous tourmaline make me pretty & sexy agate make me good, better, best amethyst make me clairvoyant malachite make me positive & passionate lazuli make me love & loved moonstone make me feel feminine & mysterious amber make me as good as new jade make me confident & happy pearl make me weep & forget citrine make me warm & tender chalcedony make me dream & fantasize chainchains. charlottewooning

aventurine make me adventurous tourmaline make me pretty & sexy agate make me good, better, best amethyst make me clairvoyant malachite make me positive & passionate lazuli make me love & loved moonstone make me feel feminine & mysterious amber make me as good as new jade make me confident & happy pearl make me weep & forget citrine make me warm & tender chalcedony make me dream & fantasize chainchains. charlottewooning

aventurine make me adventurous tourmaline make me pretty & sexy agate make me good, better, best amethyst make me clairvoyant malachite make me positive & passionate lazuli make me love & loved moonstone make me feel feminine & mysterious amber make me as good as new jade make me confident & happy pearl make me weep & forget chalcedony make me dream & fantasize chainchains. charlottewooning

aventurine make me adventurous tourmaline make me pretty & sexy agate make me good, better, best amethyst make me clairvoyant malachite make me positive & passionate lazuli make me love & loved moonstone make me feel feminine & mysterious amber make me as good as new jade make me confident & happy pearl make me weep & forget citrine make me warm & tender chalcedony make me dream & fantasize chainchains. charlottewooning

aventurine make me adventurous tourmaline make me pretty & sexy agate make me good, better, best amethyst make me clairvoyant malachite make me positive & passionate moonstone make me feel feminine & mysterious amber make me as good as new jade make me confident & happy pearl make me weep & forget citrine make me warm & tender chalcedony make me dream & fantasize chainchains. charlottewooning

aventurine make me adventurous tourmaline make me pretty & sexy agate make me good, better, best amethyst make me clairvoyant malachite make me positive & passionate lazuli make me love & loved moonstone make me feel feminine & mysterious amber make me as good as new jade make me confident & happy make me weep & forget citrine make me warm & tender chalcedony make me dream & fantasize chainchains. charlottewooning

aventurine make me adventurous tourmaline make me pretty & sexy amethyst make me clairvoyant malachite make me positive & passionate lazuli make me love & loved moonstone make me feel feminine & mysterious amber make me as good as new jade make me confident & happy pearl make me weep & forget citrine make me warm & tender chalcedony make me dream & fantasize chainchains. charlottewooning

aventurine make me adventurous tourmaline make me pretty & sexy agate make me good, better, best amethyst make me clairvoyant malachite make me positive & passionate lazuli make me love & loved moonstone make me feel feminine & mysterious jade make me confident & happy pearl make me weep & forget citrine make me warm & tender chalcedony make me dream & fantasize chainchains. charlottewooning

aventurine make me adventurous tourmaline make me pretty & sexy agate make me good, better, best amethyst make me clairvoyant malachite make me positive & passionate lazuli make me love & loved moonstone make me feel feminine & mysterious amber make me as good as new jade make me confident & happy pearl make me weep & forget citrine make me warm & tender chainchains. charlottewooning

Design by Marjolein Delhaas for Charlotte Wooning.

The redesign of the graphic image for Mat∗ters updates the product (mats) and synchronizes it with the new aesthetic trends, which tend towards minimalism and austerity rather than towards baroque. One single color and a simple symbol (the asterisk) are sufficient.

Hello Monday for Mat∗ters.

In this case, the objective is to transform the perception of bicycles from a simple mean of transportation to into a fashion accessory. This explains the use of identifying tags with the brand's logo, conceived as if they were stylish "tattoos" for the product.

Design by Hello Monday for Von Backhaus.

Book is a collection of notebooks, notepads and calendars designed by Marjolein Delhaas, which she stitches herself. Labels (the font is Akzidenz-Grotesk) allow the owner to write a title or a note both in the cover and the spine of the notebooks.

Design for the Book collection of hand-made books by Marjolein Delhaas.

The graphic pattern used on the labels, the stickers, the boxes and the bags for joid'art is one by means of the digital manipulation and saturation of the scanning of one of the pieces of jewelry from the collection. The resulting abstract image unifies the brand's image and the product's.

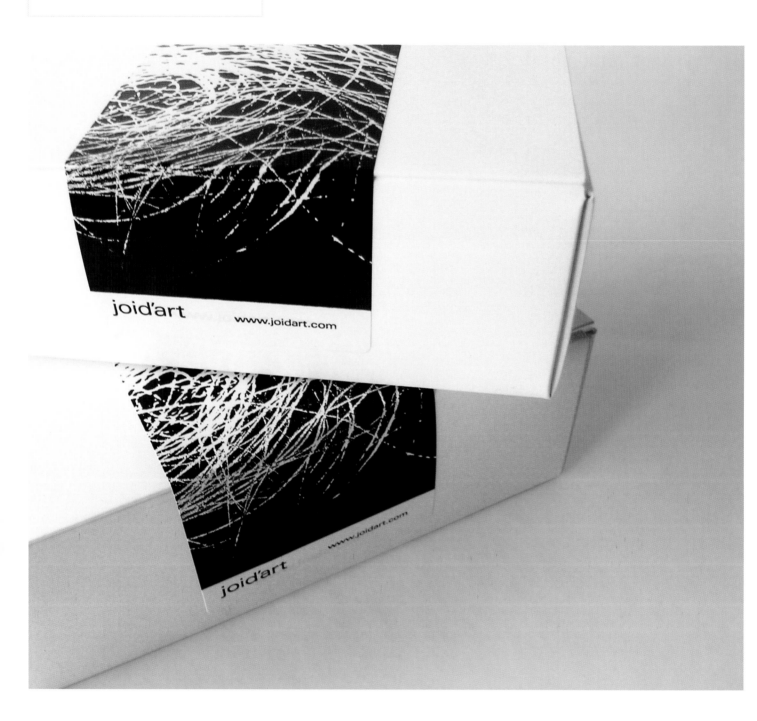

Design by Cristina Julià and the creative studio of joid'art
for the jewelry stores joid'art specializing in silver designs.

This series of six labels have specifically been designed as a closing element of envelopes and packages. They include the client's information, Marcel van Doorn and are decorated with a color dotted pattern, which forms an abstract and irregular design. The font used is Avenir Black.

Closing labels designed by Marjolein Delhaas
for the creative director Marcel van Doorn.

The dot pattern acts as a frame for the informative text printed on the label and as a decorative element on other occasions (the central rectangle of the tag on the left, for instance). To avoid the visual monotony the position of the texts are frequently changed.

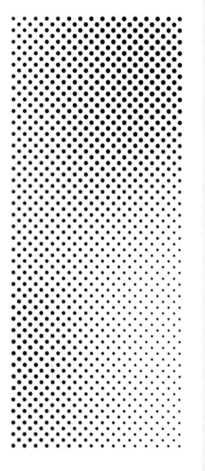

TO ADDRESS ABOVE
IF NOT DELIVERABLE PLEASE RETURN

RIJNSTRAAT 19/1
1078 PV AMSTERDAM
THE NETHERLANDS

MARCEL VAN DOORN
CREATIVE DIRECTION

TO ADDRESS ABOVE
IF NOT DELIVERABLE PLEASE RETURN

THE NETHERLANDS
1078 PV AMSTERDAM
RIJNSTRAAT 19/1

MARCEL VAN DOORN
CREATIVE DIRECTION

IF NOT DELIVERABLE PLEASE RETURN
TO ADDRESS ABOVE

Labels designed by Marjolein Delhaas for Marcel van Doorn.

Earl Grey, English Breakfast and Verveine are three of the products sold at Birgit Israel's design store. The labels show a different color for each type of tea, but unifies the design thanks to the product's logo and the romantic style frieze that surrounds the central label.

Boxes and labels designed by Richard Ardagh for Brigit Israel.

The graphic image of Jackie Smith has been developed based on the concept of "accessible luxury" starting with the design of the brand's visual identity. The packaging has been set out so the brand would be slowly revealed, surprising with every design detail.

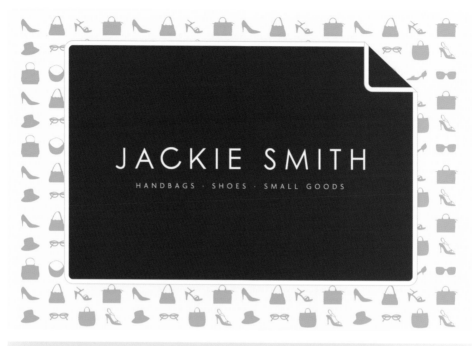

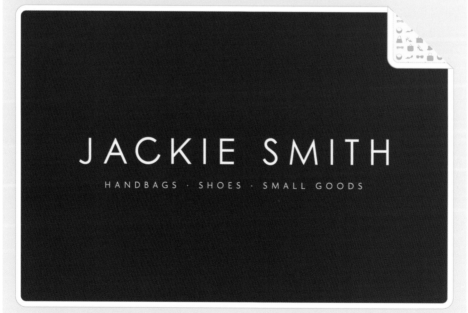

Designs by Fbdi for Jackie Smith.

For the signage of Lomo's Parisian store, adhesive labels have been printed and later mounted on PVC (polyvinyl chloride) rigid panels. The aesthetics is inspired on the graphics in airports and on commercial airplanes.

Labels and signage designed by Lomographic Society International for Lomography Gallery Shop Paris.

Boxes designed by Lomographic Society International
for Lomography Gallery Shop Tokyo.

This set of labels for Lomo designed after the shape of a camera, and the brand's logos printed on transparent paper to be used on any kind of surface. The labels act as the brand's promotional cards and can be collectable.

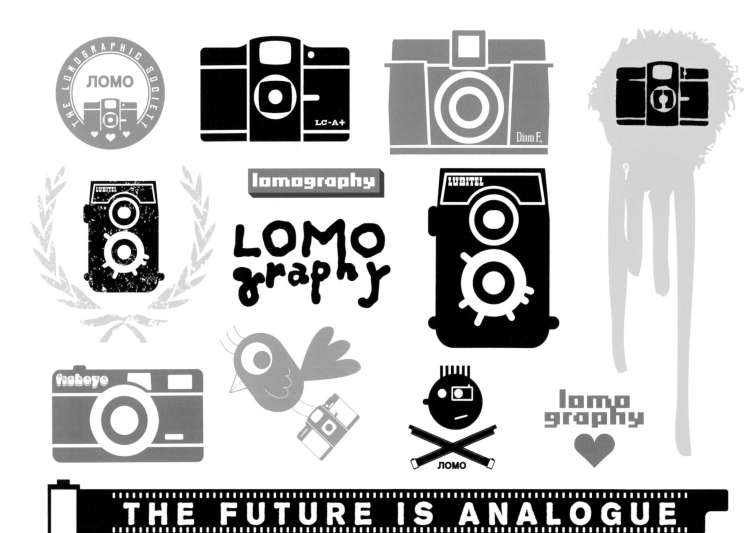

Set of stickers, 11 × 17 inches designed by Lomographic Society International.

SUPERSAMPLER

One push on the shutter release button = four sequential panoramic pictures in a row on one photo print. Lightning fast film advance with patented rip-cord system.

一度シャッターをきれば、ワンショットで1枚の写真の中に4つの連続したパノラマ・イメージが順番に並びます。プルコードをさっと引っ張るだけのかんたんフィルム送りで、素早く次のシューティングへ。

SUPERSAMPLER

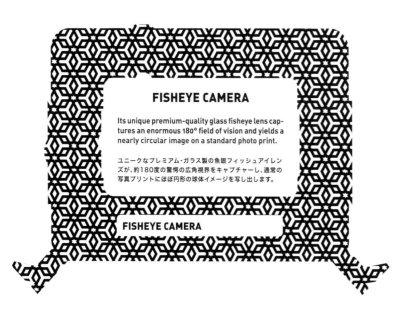

FISHEYE CAMERA

Its unique premium-quality glass fisheye lens captures an enormous 180° field of vision and yields a nearly circular image on a standard photo print.

ユニークなプレミアム・ガラス製の魚眼フィッシュアイレンズが、約180度の驚愕の広角視界をキャプチャーし、通常の写真プリントにほぼ円形の球体イメージを写し出します。

FISHEYE CAMERA

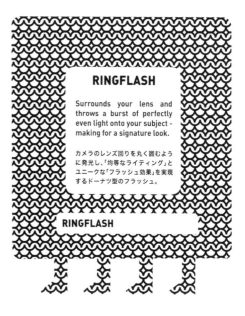

RINGFLASH

Surrounds your lens and throws a burst of perfectly even light onto your subject – making for a signature look.

カメラのレンズ回りを丸く囲むように発光し、「均等なライティング」とユニークな「フラッシュ効果」を実現するドーナツ型のフラッシュ。

RINGFLASH

Labels designed by Lomographic Society International.

Labels

The labels' design for the boxes of the earphone fabricator Hulger intends to avoid a quick obsolescence of the device betting for a modern but timeless graphic design in the same way the products of the brand avoid the technologies condemned to go out of style.

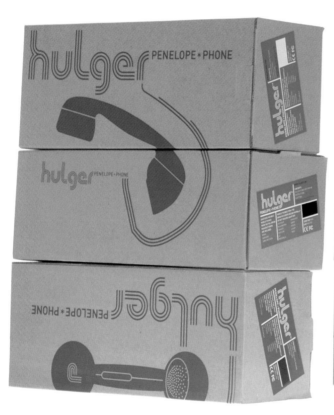

Design by Hulger. Graphic design by Violetta Boxhill. Product design by Nicolas Roope and Kam Young.

Orbiculus is a complete set of 10 pins designed by Art. Lebedev Studio inspired on the traditional commands of computer language. Consequently, the label shows the typical hand cursor that can be seen on a computer screen.

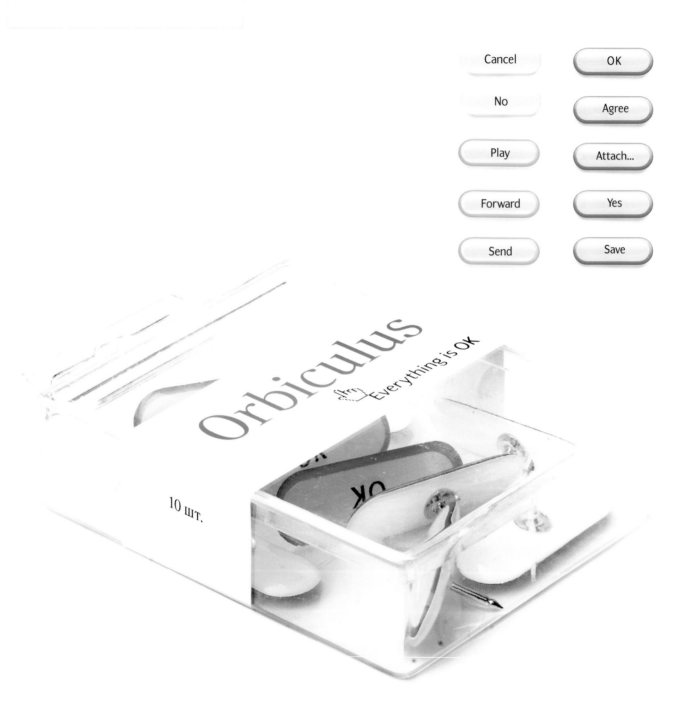

Design by Yuri Shvetsov and Roma Voroneshski of Art Lebedev Studio.

Boxes designed by Pablo Martín, Meri Iannuzzi, Borja Martínez
and Mónica Llena of grafica for Metalarte.

The design of these labels achieves a simple requirement: show a graphic image of the low cost high-end line of products. This explains the presentation of the information in the form of an instructions manual or patient information leaflet.

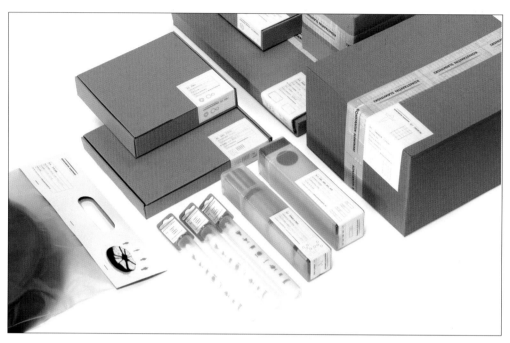

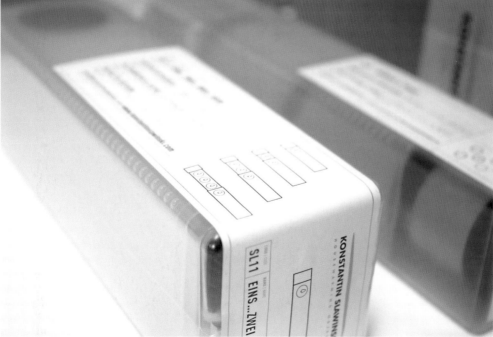

Packaging and adhesive labels designed by Christopher Ledwig of F1RSTDESIGN.com
for Konstantin Slawinski's collection of objects for the home.

Labels

These sticker-labels show the basic human emotions (boredom, happiness, humiliation…) and imitate the format of the periodic table of chemical elements. The labels are used to close the envelopes and the bags of the products for the brand Test Tube.

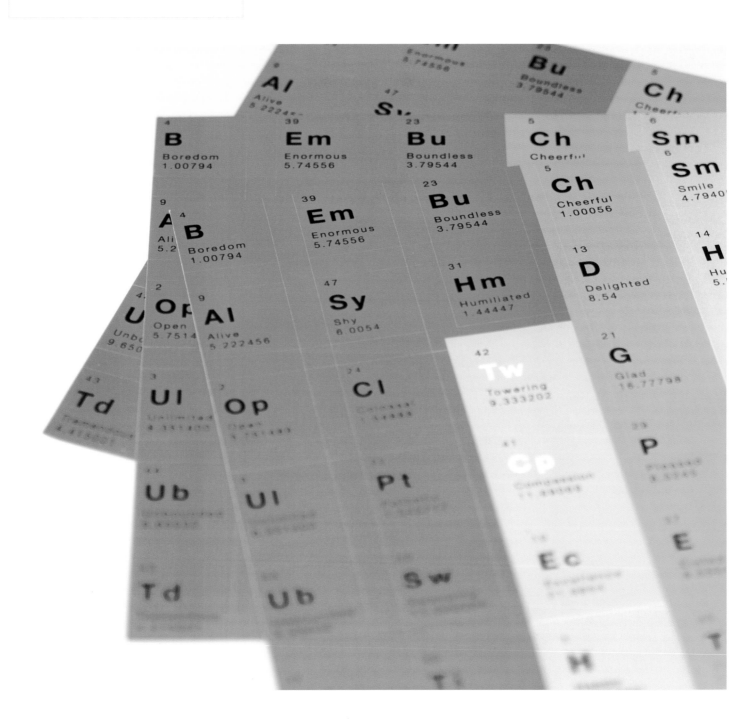

3/4 × 3/4 inch labels printed in three colors and designed by Block for the design store Test Tube in Perth (Australia).

Boxes designed by Pablo Martín and Ellen Diedrich of grafica
for Ready-Made, Janina Bea and Elena Mateu Pomar.

The logo for M… TheMovement has been engraved with a burin on khaki and charcoal grey patches leather made to look old and also on the brand's denim garments. The font was specifically created by the designer for M… TheMovement.

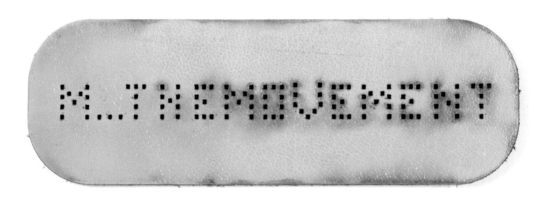

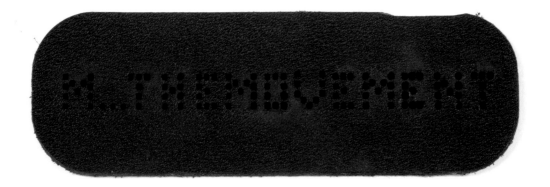

The hangers designed by Gary Barker go beyond their obvious practical function and work also as promotional elements. This explains the printed logo of the brand and the distinctive illustration: two human figures within an ornamented crown.

Recycled cardboard patches and clothes rack designed by Gary Barker of Greenheart Global for M… TheMovement.

The embroidered black labels come with the originally customized garments (cut and sew) for Mishka NYC, while the pink label is reserved for the T-shirts, marking the difference between collections. The use of the Cyrillic alphabet is the brand's wink to its own name.

Design by Mishka NYC.

The tags for the Swiss fashion house +alprausch+ always show the silhouette of two animals challenging each other (deer, swans, or lions) on a background of radiating lines that emanate from the center, similar to how the solar rays might radiate.

Labels designed by Thomas Brader of Komun GmbH for the Swiss brand +alprausch+.

The cardstock labels for IVANAhelsinki with the red logo are included in the products printed with patterns and other hand-made details. Unlike the rest of the brand's labels, they do not present any other information on their back.

IVANAhelsinki ✦

IVANAHELSINKIHemma

IVANAHELSINKIHemma

100 % COTTON

WASH INSIDE OUT
WITH SIMILAR
COLORS
RESHAPE AND
STRETCH INTO
SHAPE IMMEDIA-
TELY AFTER GENTLE
WASH WHILE
GARMENT IS STILL
MOIST
BY IRONING THE
MOIST GARMENT
YOU CAN RETAIN ITS
ORIGINAL SHAPE
SOME SHRINKAGE
MAY OCCUR

36

IVANAhelsinki ✦

Homemade in Finland
Did you know that you
can also take care of your
clothes by airing, brushing
and storing them well

38

IVANAhelsinki ✦

Homemade in Finland
Did you know that you
can also take care of your
clothes by airing, brushing
and storing them well

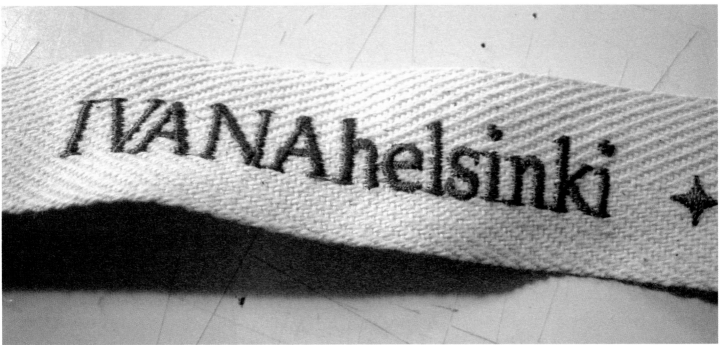

Designs by Paola Ivana Suhonen for IVANAhelsinki.

Labels

The collaboration between Lars P. Amundsen with Siv Støldal started in 2002 as a favor, when the first did a cover for lookbook in which he included the characteristic diagonal line across the O. Since then, this diacritical mark has become the brand's identity symbol.

Design by Lars P. Amundsen for Siv Støldal.

Siv Støldal's labels do not contain any type of additional information other than the brand's logo in red, the season to which the garment belongs (the image on this page, fall-winter 2006-2007) and the size.

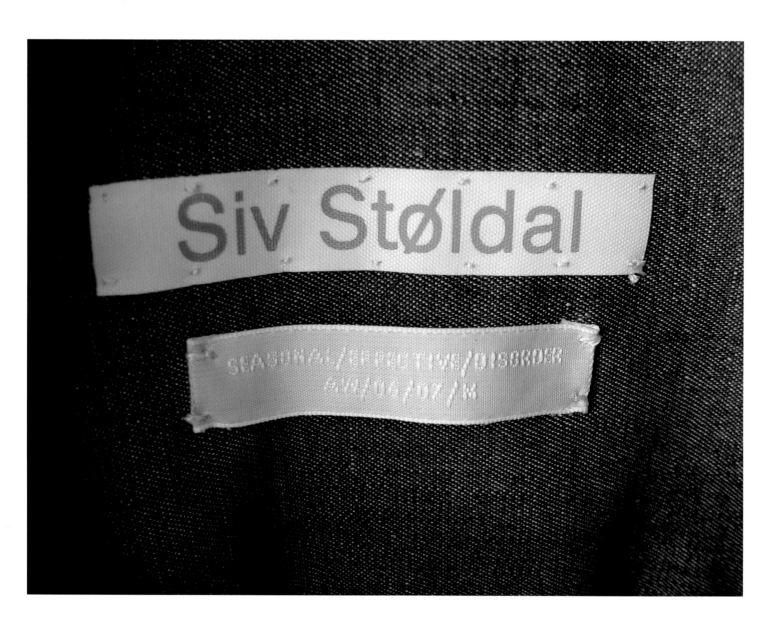

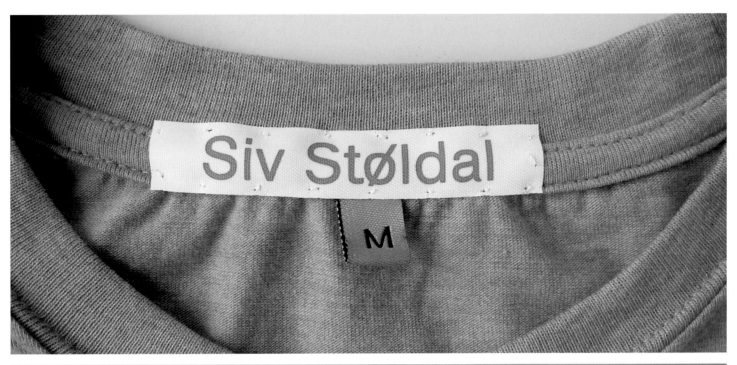

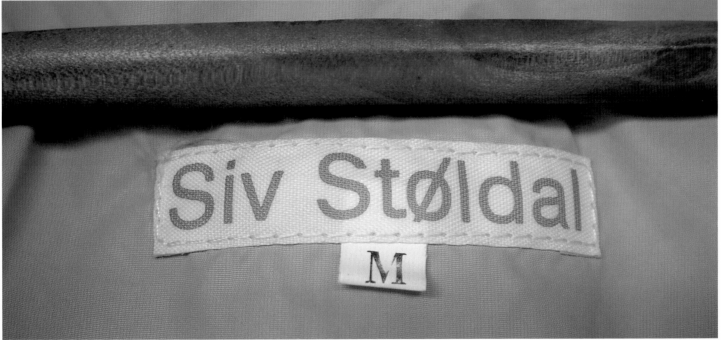

Designs by Lars P. Amundsen for Siv Støldal.

Adieu Tristesse is a Japanese fashion brand. A simple design of traditional European flavor was chosen for the garments' labels. The label for Les touristes belongs to the suitcase collection designed in collaboration with Atelier LZC.

Top: design by Atelier LZC for Adieu Tristesse. / Below: design by
Atelier LZC and Les touristes for Les touristes.

Design by Atelier LZC for Tissage Moutet.

In this case, it is an "author" T-shirt that can be customized with the client's written messages, and the label has to provide all the information about the design and the name of the project, although without competing or taking away importance from the product.

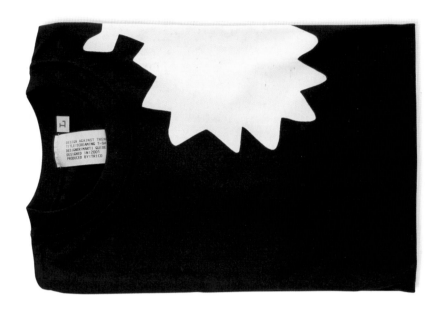

Martí Guixé for Trico. © of the images: IMAGEKONTAINER.

On this page: the main logo for the brand Katty Xiomara (awarded with Excellence Certificate by the European Logo Annual Design 1999) is on all the brand's labels. On the next page: the logo for the Intimo line by Katty Xiomara is radically different from the brand's main logo.

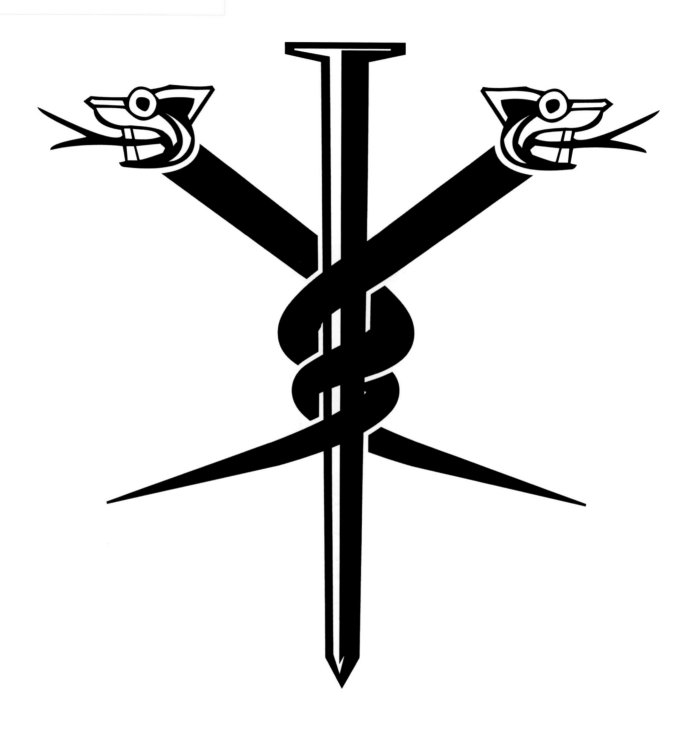

Design by Katty Xiomara and José M. Da Silva for Katty Xiomara.

The different logos for Katty Xiomara, as well as the variants of each of them appear both on the labels of the brand's different lines and on the buttons of the garments. Also, as pattern on the wrapping paper.

Designs by Katty Xiomara and José M. Da Silva for Katty Xiomara.

The labels for Ooito Japan show different traditional Japanese graphic icons and designs inspired on the gundam culture. The label on this page has been silk screened on fabric with one ink. The tags, which have been sown, are multicolor.

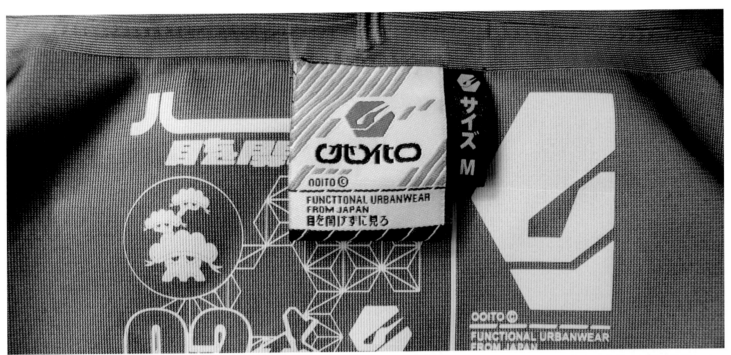

Labels designed by Sonia Chow & Huschang Pourian
of ChowPourianLab for Ooito Japan.

The name for the ski jackets line by Ooito Japan, Double Fuji, does not make reference to Mount Fuji but to the kanji (Chinese characters) for "rich" and "warrior." This explains the selection of the kabuto helmet as icon.

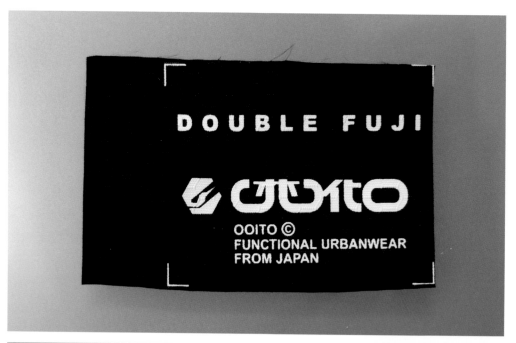

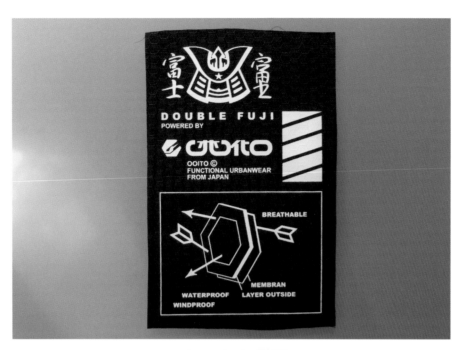

Logo and the label for the lining of the ski jackets by Ooito Japan.
Design by Sonia Chow & Huschang Pourian.

To avoid falling in the "generic" design of most labels, ChowPourianLab personalized those for Ooito with 13 characters each with a strong personality. Jett (top) is a judo champion. Ginger (below) is a handyman capable to fix anything.

Labels for T-shirts. Design by Sonia Chow & Huschang Pourian
of ChowPourianLab for Ooito Japan.

To avoid the frequent irritation and allergy problems that the conventional labels cause on some people, Alpinestars has printed the labels directly on the inside of the garment. The result is ornamental.

Labels for the main clothing line and the Nero line by Alpinestars.

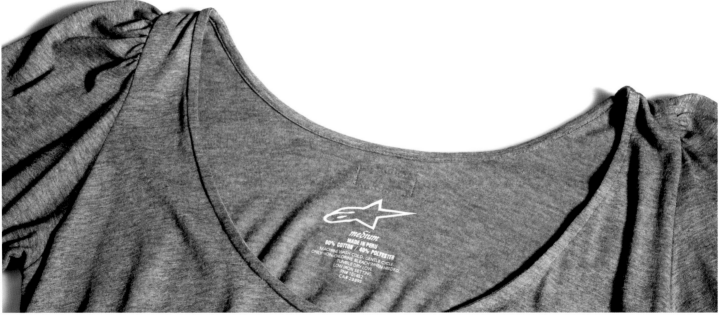

Design by the Alpinestars team.

The decorative patches and the labels for Freshjive have been fabricated with unconventional materials (such as leather for instance). On this page, the logo, coarsely embroidered brings in a striking craft detail.

Design by Freshjive (front and back).

The decorative patches used in some Freshjive garments work simultaneously as the product's identifying "exterior" label and as hallmark. The iconic images are repeated on all the elements and help capturing the client's attention.

Design by Freshjive.

For the Agua Bendita labels, a graphic language was decided that makes reference to the "old school" tattoos for the character of the line and chosen colors reflecting the vintage aesthetics of the swimwear collection Aquatic Cosmos.

Design by Daniel Mejía (design) and Julian Román (illustrations)
of Plasma for the swimwear brand Agua Bendita.

The vertical stripe pattern of different thicknesses and colors unifies the design of the label for the brand BB Dakota. The typography of the informative text has a calligraphy aesthetic and provides the design with a warm detail while it aims to sympathize with the client.

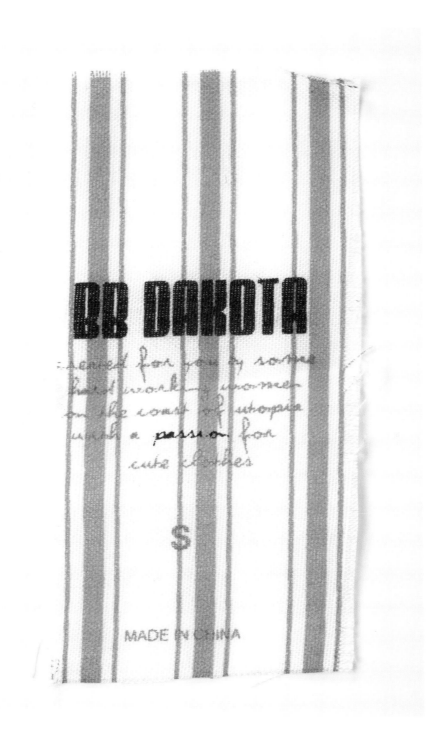

Designs by BB Dakota.

A perforated metal sheet decorated with a graphic detail serves to hold the piece of fabric onto which the brand's logo has been printed. On the next page: both fabric label are sown next to the spare button for the garment.

Designs by BB Dakota.

The bright fluorescent colors provide BB Dakota's label with their own identity and transform them into a decorative object, regardless of the garment to which they are attached.

2 2 3 2 5 6 8 2

Designs by BB Dakota.

The heraldry inspired graphics based on the coat of arms have frequently been used by the urban fashion brands in recent time. The trend has also arrived to the design of the label. In this case, BB Dakota has inserted its logo in a heraldic flag.

Designs by BB Dakota.

The ribbons of fabric sown to the labels are a decorative element. The goal is that the label functions as an ornamental piece for the garment and not just as a functional element. This explains the use of brightly colored thread in the stitching of the label.

Designs by BB Dakota.

The label for the Berlin-based brand of urban fashion IRIEDAILY, created in 1994, has the objective
to convey the irreverent, fresh, urban spirit strong-
ly connected to the art world, which the brand's
graphic design is inspired by.

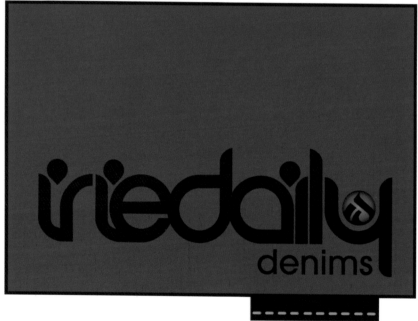

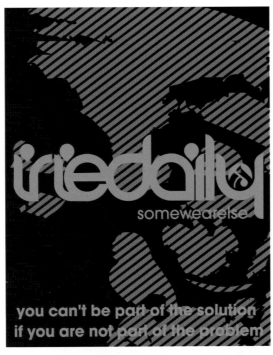

Designs by Jaybo aka Monk of IRIEDAILY.

Fabricated with high definition nylon, the label on this page belong to SKUNKFUNK's spring-summer 2008 collection. The goal was to use geometric shapes as well as natural ones to make reference to summer.

Designs by Susana Sánchez Monje and Sergio Llanos of SKUNKFUNK.

SKUNKFUNK's labels often have different designs depending on whether the garment they come with is from the men or women's collection. On this double page, the labels are fabricated in nylon for the women's collection and make reference to the fifties in Cuba, the sea, the wind and music.

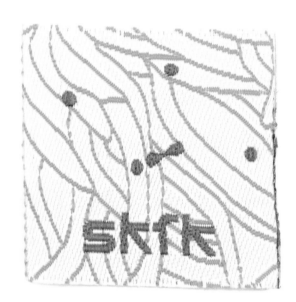

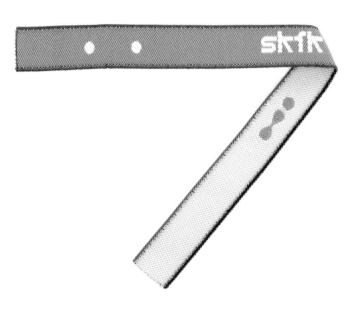

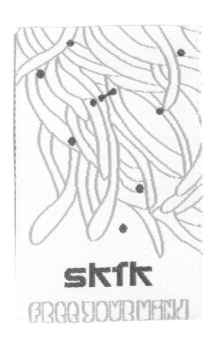

Designs by Susana Sánchez Monje and Sergio Llanos of SKUNKFUNK.

With the idea in mind that the brand's graphic image of the label should be anything but monotonous, SKUNKFUNK has played, in this case, with the negatives of its designs. The brand's logo, however, is always embroidered or printed in the green corporative color.

Designs by Susana Sánchez Monje and Sergio Llanos of SKUNKFUNK.

The abstract graphic pattern embroidered shown on this page is repeated on various labels for the spring-summer 2009 collection. The design is inspired by three concepts: "rainbow," "color" and "funk."

Designs by Susana Sánchez Monje and Sergio Llanos of SKUNKFUNK.

The image's thick cardboard band has been printed in one ink. The fabric label shows the brand's logo embroidered on its front and the web page on its back. The logo is always printed in black.

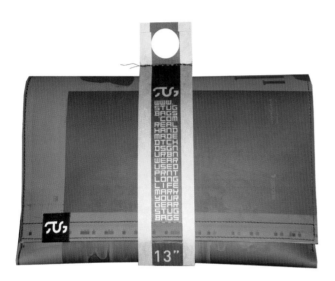

Fabian Meijer and Pim Nap of Studio Piraat for STUGBAGS.

Dan Jonsson and Louise Hederström by Charming Unit for Byggfabriken.

The illustrations embroidered on the tag for the Argentinean brand Chicas Pin Ups have been hand-drawn by the designer. The font is Luxury, while the two colors (pale pink and fluorescent green) fit in the pop and adolescent aesthetic of the firm.

Label designed by Mek Frinchaboy for Chicas Pin Ups.

Label designed by Mek Frinchaboy for Think Pink.

Originally, the audio tape image embroidered on the Lazy Oaf label was going to be a graphic element that would reflect the illustrations of the brand. However, it ended up becoming one of the brand's identifying symbols.

Design by Gemma Shiel of Lazy Oaf.

Inspired by the work of Otto Neurath, Gerd Arntz, and Rudolf Modley, Charming Unit designed the tags for Byggfabriken with added pictograms organized in a pattern of typical situations that can be experienced in the stores of the Swedish brand.

BYGGFABRIKEN

LINOLJA

KOKT KALLPRESSAD

FRÅN VIRESTADSGÅRDEN

REN OCH LJUS LINOLJA SOM TORKAR GENOM OXIDERING.

KOKT LINOLJA ÄR MER TRÖGFLYTANDE OCH HAR KORTARE TORKTIDEN RÅ LINOLJA. LÄMPLIG TILL YTBEHANDLING AV INVÄNDIGA YTOR, SOM BINDEMEDEL I LINOLJEFÄRG SAMT ROSTSKYDD TILL PLÅT OCH SMIDE.

YTAN BEHANDLAS UPPREPADE GÅNGER TILLS DEN ÄR MÄTTAD. LINOLJA SOM INTE ABSORBERAS INOM 1–2 TIMMAR TORKAS AV. LÅT TORKA ORDENTLIGT INNAN TÄCKMÅLNING. BALSAMTERPENTIN TILLSÄTTS FÖR ATT FÖRKORTA TORKTIDEN.

VARNING! LINOLJEINDRÄNKTA TRASOR KAN SJÄLVANTÄNDA.

WWW.BYGGFABRIKEN.COM

BYGGFABRIKEN

LINOLJA

RÅ KALLPRESSAD

FRÅN VIRESTADSGÅRDEN

REN OCH LJUS LINOLJA SOM TORKAR GENOM OXIDERING.

RÅ LINOLJA ÄR TUNNFLYTANDE OCH HAR GOD INTRÄNGNINGS-FÖRMÅGA I TRÄ OCH ÄR DÄRFÖR LÄMPLIG TILL IMPREGNERING AV UTVÄNDIGA YTOR, SOM BINDEMEDEL I LINOLJEFÄRG SAMT ROSTSKYDD TILL PLÅT OCH SMIDE.

YTAN BEHANDLAS UPPREPADE GÅNGER TILLS DEN ÄR MÄTTAD. LINOLJA SOM INTE ABSORBERAS INOM 1–2 TIMMAR TORKAS AV. LÅT TORKA ORDENTLIGT INNAN TÄCKMÅLNING. BALSAMTERPENTIN TILLSÄTTS FÖR ATT FÖRKORTA TORKTIDEN.

VARNING! LINOLJEINDRÄNKTA TRASOR KAN SJÄLVANTÄNDA.

WWW.BYGGFABRIKEN.COM

Dan Jonsson and Louise Herderström of Charming Unit for Byggfabriken.

Directory

29 Agency
www.29agency.com
info@29agency.com

Afroditi Krassa
www.afroditi.com
krassa@afroditi.com

Alpinestars
www.alpinestars.com
mediaservice3@alpinestars.com

Art. Lebedev Studio
www.artlebedev.com
mailbox@artlebedev.com

Atelier LZC
www.atelierlzc.fr
celine@atelierlzc.fr

BB Dakota
www.bbdakota.net
jacquelyn.morell@bbdakota.com

Block
www.blockbranding.com
someone@blockbranding.com

Boost
www.boostmobile.com
press@boostmobile.com

Bruketa&Zinic
www.bruketa-zinic.com
bruketa-zinic@bruketa-zinic.com

Charming Unit
www.charmingunit.com
hello@charmingunit.com

ChowPourianLab
www.chowpourian.com
contact@chowpourian.com

Curious Design Consultants
www.curious.com.au
design@curious.co.nz

Curse of the Multiples
www.curseofthemultiples.com
info@curseofthemultiples.com

David Barath
www.davidbarath.com
david@visualgroup.hu

Dil Brands
www.dilbrands.com
fmuniz@dilbrands.com

EIGHT
www.8design.jp
info@8design.jp

F1RSTDESIGN.com
www.f1rstdesign.com
hello@f1rstdesign.com

Fbdi
www.estudiofbdi.com
info@estudiofbdi.com

**Feldmann+Schultchen
Design Studios**
www.fsdesign.de
mail@fsdesign.de

Freshjive
www.freshjive.com
contact@freshjive.net

Fujizaki
www.fujizaki.com
contact@fujizaki.com

Gaia Bisconti Design
gaia.bisconti@gmail.com

grafica
www.grafica-design.com
grafica@grafica-design.com

HakGraphics
www.hakgraphics.com
info@hakgraphics.com

Haus of M
www.hausofm.com
m@hausofm.com

Hello Monday
www.hellomonday.net
you_had_me@hellomonday.net

Hulger
www.hulger.com
info@hulger.com

IRIEDAILY
www.iriedaily.de
info@iriedaily.de

IVANAhelsinki
www.ivanahelsinki.com
pirjo@ivanahelsinki.com

JOED Design
www.joeddesign.com
erebek@joeddesign.com

joid'art
www.joidart.com
joidart@joidart.com

KanaBeach
www.kanabeach.com
international@kanabeach.com

Katty Xiomara
www.kattyxiomara.com
info@kattyxiomara.com

Ketchup&Majo
www.ketchupandmajo.com
info@ketchupandmajo.com

KGB+FOX
www.kgb-free.com
hello@kgb-free.com

Komun GmbH
www.komun.ch
grafik@komun.ch

Kulte
www.kulte.fr
nico@kulte.fr

Lars P. Amundsen
www.designbylars.com
info@designbylars.com

Lazy Oaf
www.lazyoaf.com
gemma@lazyoaf.com

Lomographic Society International
www.lomography.com
design@lomography.com

Mads Nørgaard
www.madsnorgaard.com
info@madsnorgaard.dk

Malax Design
www.malax-design.com
thomas@malax-design.com

Marjolein Delhaas
www.marjoleindelhaas.com
info@marjoleindelhaas.com

Martí Guixé
www.guixe.com
info@guixe.com

Mek Frinchaboy
www.flickr.com/photos/frinchagirl
mekfrinchaboy@gmail.com

Mishka NYC
www.mishkanyc.com
info@mishkanyc.com

Morera Design
www.moreradesign.com
morera@moreradesign.com

Onda
www.ondaworld.com
mdmarcas@gmail.com

Optic Garment
www.opticgarment.com
optic@opticgarment.com

Optima Soulsight
www.optimasoulsight.com
info@optimasoulsight.com

Paristanbul
www.paristanbul.net
info@paristanbul.net

Pearlfisher
www.pearlfisher.com
briefus@pearlfisher.com

Plasma
www.plasma4.com
contactos@plasma4.com

Prime Studio
www.primestudio.com
stuart@primestudio.com

Richard Ardagh
www.elephantsgraveyard.co.uk
info@elephantsgraveyard.co.uk

Setsu & Shinobu Ito
www.studioito.com
studioito@studioito.com

SiO DESIGN
www.sio-design.co.jp
info@sio-design.co.jp

SKUNKFUNK
www.skunkfunk.com
info@skunkfunk.com

The Smiley Company
www.smileylicensing.com
kyung@smileyworld.co.uk

SMR Creative
www.smrcreative.com
studio@smrcreative.com

Staynice
www.staynice.nl
staynice@staynice.nl

Stefanie Reeb
www.stefaniereeb.de
mail@stefaniereeb.de

Stormhand
www.stormhand.com
boy@stormhand.com

Studio Piraat
www.studiopiraat.nl
info@studiopiraat.nl

Swerve
www.swerveinc.com
contact@swerveinc.com

The Creative Method
www.thecreativemethod.com
mail@thecreativemethod.com

Tomasoni Topsail
www.tomasoni.com
group@tomasoni.com

Visual Group
www.visualgroup.hu
info@visualgroup.hu

Yes No Maybe
www.yesnomaybe.co.uk
info@yesnomaybe.co.uk

Zwei
www.zwei-bags.com
info@zwei-bags.com